**Pauli's Exclusion Principle**
The Origin and Validation of a Scientific Principle

There is hardly another principle in physics with wider scope of applicability and more far-reaching consequences than Pauli's exclusion principle. This book explores the origin of the principle in the atomic spectroscopy of the early 1920s, its subsequent embedding into the emerging quantum mechanics, and the later experimental validation with the development of quantum chromodynamics and parastatistics.

The origin of the exclusion principle in 1924 is intertwined with the discovery of the electron's spin, which marked the crisis of the old quantum theory and the transition to quantum mechanics. The reconstruction of this crucial historical episode provides an excellent foil to reconsider Thomas Kuhn's view on incommensurability. In this book, Michela Massimi defends the prospective rationality of this revolutionary transition by focussing on the specific way in which Pauli's principle emerged as a phenomenological rule 'deduced' from some anomalous phenomena and theoretical assumptions of the old quantum theory. The process of validation, which took place in the following decades and transformed Pauli's rule into an important scientific principle, is analysed from both historical and philosophical points of view. A suitable version of 'dynamic Kantianism' is proposed as the philosophical framework for an understanding of the role and function of the exclusion principle.

This historico-philosophical investigation touches upon some of the most relevant issues in philosophy of science and suggests new answers. The variety of themes skilfully woven together makes this book of interest to philosophers, historians, physicists and those with an interest in philosophy working in the natural and social sciences.

MICHELA MASSIMI is a Research Fellow at Girton College, University of Cambridge, affiliated with the Department of History and Philosophy of Science. She gained her *Laurea* in Philosophy at the University of Rome 'La Sapienza' in 1997. Her *Laurea* thesis on the Bohr–Einstein debate on the completeness of quantum mechanics won the Prize of the Accademia Nazionale delle Scienze, detta dei XL in 1998. She gained an M.Phil./Ph.D. from the London School of Economics following research in fields including history and philosophy of science, epistemology, scientific methods and the history of quantum mechanics.

Dr Massimi has lectured on philosophy of science and philosophy of physics courses at the University of Cambridge, at the Philosophy Faculty and at the Physics Department, Cavendish Laboratory. From October 2005 she is Lecturer in History and Philosophy of Science at the Department of Science and Technology Studies of University College London.

# Pauli's Exclusion Principle
## The Origin and Validation of a Scientific Principle

MICHELA MASSIMI

CAMBRIDGE UNIVERSITY PRESS
Cambridge, New York, Melbourne, Madrid, Cape Town, Singapore, São Paulo

CAMBRIDGE UNIVERSITY PRESS
The Edinburgh Building, Cambridge CB2 2RU, UK

www.cambridge.org
Information on this title: www.cambridge.org/9780521839112

© M. Massimi 2005

This book is in copyright. Subject to statutory exception
and to the provisions of relevant collective licensing agreements,
no reproduction of any part may take place without
the written permission of Cambridge University Press.

First published 2005

Printed in the United Kingdom at the University Press, Cambridge

*A catalogue record for this book is available from the British Library*

ISBN-13 978-0-521-83911-2 hardback
ISBN-10 0-521-83911-4 hardback

Cambridge University Press has no responsibility for the persistence or accuracy of URLs for external or third-party internet websites referred to in this book, and does not guarantee that any content on such websites is, or will remain, accurate or appropriate.

Every effort has been made to secure necessary permissions to reproduce copyright material in this work, though in some cases it has proved impossible to trace copyright holders. If any omissions are brought to our notice, we will be happy to include appropriate acknowledgements on reprinting or in any subsequent edition.

*A Luciana e Gianni*

# Contents

|   |   |   |
|---|---|---|
| *Note on translation* | *page* | x |
| *Preface* | | xi |
| *Epigraph* | | xiv |
| **Introduction** | | 1 |
| **1** | **The exclusion principle: a philosophical overview** | **7** |
| 1.1 | Introduction | 7 |
| 1.2 | From Poincaré's conventionalism to Popper and Lakatos on the nature of the exclusion principle | 9 |
| 1.3 | From Reichenbach's coordinating principles to Friedman's relativized a priori principles | 13 |
| 1.4 | Constitutive versus regulative | 21 |
| | 1.4.1 Kant on the regulative principle of systematicity | 25 |
| | 1.4.2 Ernst Cassirer and the architectonic of scientific knowledge | 28 |
| 1.5 | The exclusion principle: a Kantian perspective | 31 |
| **2** | **The origins of the exclusion principle: an extremely natural prescriptive rule** | **35** |
| 2.1 | The prehistory of Pauli's exclusion principle | 35 |
| | 2.1.1 Atomic spectra and the Bohr–Sommerfeld theory of atomic structure | 35 |
| | 2.1.2 The doublet riddle and the riddle of statistical weights | 43 |

|   |   | 2.1.3 The anomalous Zeeman effect and the mystery of half-integral quantum numbers | 47 |
|---|---|---|---|
| 2.2 | | Bohr, Heisenberg, and Pauli on spectroscopic anomalies | 52 |
|   |   | 2.2.1 Niels Bohr: nothing but a 'non-mechanical constraint'? | 52 |
|   |   | 2.2.2 Heisenberg's first core model: the sharing principle. Does success justify the means? | 55 |
|   |   | 2.2.3 Heisenberg's second core model: the branching rule and a new quantum principle | 60 |
|   |   | 2.2.4 Pauli: from the electron's *Zweideutigkeit* to the exclusion rule | 65 |
| 2.3 | | The turning point | 73 |
| **3** | | **From the old quantum theory to the new quantum theory: reconsidering Kuhn's incommensurability** | **78** |
| 3.1 | | The revolutionary transition from the old quantum theory to the new quantum theory | 78 |
| 3.2 | | Reconsidering Kuhnian incommensurability | 81 |
|   |   | 3.2.1 Kuhn on scientific lexicons: incommensurability as untranslatability | 81 |
|   |   | 3.2.2 Kuhn's argument for untranslatability and Hacking's taxonomic solution to the new-world problem | 86 |
|   |   | 3.2.3 Lexical taxonomies: the Aristotelian tradition and the nominalist criticism | 91 |
|   |   | 3.2.4 How should we read lexical taxonomies? A Kantian reading | 93 |
|   |   | 3.2.5 Reintroducing history in scientific lexicons: a lesson from the crisis of the old quantum theory | 97 |
| 3.3 | | The prospective intelligibility of the revolutionary transition from the atomic core model to the electron's *Zweideutigkeit* | 103 |
|   |   | 3.3.1 The electron's *Zweideutigkeit* and Pauli's exclusion rule as the conclusions of two nested demonstrative inductions | 103 |
| **4** | | **How Pauli's rule became the exclusion principle: from Fermi–Dirac statistics to the spin–statistics theorem** | **112** |
| 4.1 | | Introduction | 112 |
| 4.2 | | Pauli's rule prescribes a new exclusion: Fermi–Dirac statistics | 115 |
| 4.3 | | The non-relativistic quantum mechanics of the magnetic electron: Pauli's spin matrices | 119 |
| 4.4 | | Group theory enters the scene | 122 |

| | | |
|---|---|---|
| 4.5 | From quantum electrodynamics to quantum field theory: the exclusion principle re-expressed in terms of anticommutation relations | 123 |
| 4.6 | Towards relativistic quantum mechanics: the Dirac equation for the electron and the hole theory | 128 |
| 4.7 | Pauli against the hole theory: the Pauli–Weisskopf 'anti-Dirac' paper | 133 |
| 4.8 | Pauli's first proof of the spin–statistics theorem | 138 |
| 4.9 | Pauli's final proof of the spin–statistics theorem | 138 |
| 4.10 | How Pauli's rule gained the status of a scientific principle | 141 |
| **5** | **The exclusion principle opens up new avenues: from the eightfold way to quantum chromodynamics** | **145** |
| 5.1 | Introduction | 145 |
| 5.2 | From the eightfold way to quarks | 147 |
| 5.3 | Revoking or retaining the exclusion principle? | 154 |
| | 5.3.1 Revoking the strict validity of the exclusion principle: quarks as parafermions | 154 |
| | 5.3.2 Retaining the exclusion principle: coloured quarks and quantum chromodynamics | 162 |
| 5.4 | The Duhem–Quine thesis: epistemological holism and the validation of the exclusion principle | 172 |
| | 5.4.1 The validating role of negative evidence | 175 |
| | 5.4.2 Quinean underdetermination and the rationality of retaining a threatened principle | 179 |
| | **Conclusion** | **184** |
| | *References* | 189 |
| | *Index* | 204 |

# Note on Translation

Most of the historical sources quoted in this book are written in German, and in most cases no English translation is available. Thus the translations in English are my own, unless otherwise indicated in the footnotes.

# Preface

This book is the result of almost ten years of research. It has accompanied me through an intense period of my life, from the end of my undergraduate studies in Rome across the years of my Ph.D. in London until my current Research Fellowship at Girton College (University of Cambridge, UK). I have grown with it, and with it, I have come to develop my philosophical ideas. Looking back, I can see the way they have evolved and focussed; how they came to be refined, and sometimes revised. I owe intellectual debts to many people who in various ways have contributed to the development of my ideas over this span.

My original intention of studying the exclusion principle dates back to 1996. At that time I was an undergraduate student in Rome, very keen on philosophy of science and history of modern physics. Reading Pauli's scientific correspondence, I was struck by a passage of a letter to Landé in which the famous exclusion principle was introduced as an 'extremely natural rule'. It may have appeared 'extremely natural' to Pauli, but to me the overall manoeuvre seemed mysterious and intriguing. I could not help plunging into the details of this fascinating historical episode. I owe an old debt to my teachers Silvano Tagliagambe, who hooked me on philosophy of science, and Sandro Petruccioli, who encouraged me to consider Wolfgang Pauli as a possible research topic.

During the stimulating years of my Ph.D. at the London School of Economics, my research project received a new twist. It became clear how the history of the exclusion principle was intertwined with some crucial philosophical issues, such as the nature of scientific principles, the rationality of theory-choice, the underdetermination of theory by evidence as well as more specific topics in philosophy of physics such as the spin–statistics theorem. I am very grateful to my Ph.D. supervisor, Michael Redhead, who introduced me to philosophy of physics and encouraged me to work on the history of the proof of the spin–statistics theorem. Our many discussions together, his help and patient guidance during these years have been crucial for an understanding of the issues presented here. Sections 4.6–4.9 are developed from our joint article 'Weinberg's proof of the spin–statistics theorem', *Studies in History and Philosophy of Modern Physics* **34** (2003), 621–50 (Copyright (2003) by Elsevier. Reprinted with kind permission from Elsevier). I am very grateful also to my Ph.D. co-supervisor Carl Hoefer: his constructive and friendly criticism helped me clarify some philosophical points and better articulate my views.

The staff of the Science Museum Library in London showed great patience in dealing with my request for microfilms from the Archive for the History of Quantum Physics. I would like to thank also the Department of Philosophy, Logic and Scientific Method at the LSE as well as the Arts and Humanities Research Board (AHRB) for financial support during the years of my Ph.D.

This book builds upon my Ph.D. thesis, yet it has ended up being quite different and distinct from it. The past three years at Cambridge have been most fruitful and inspiring for refining my philosophical view. I thank first and foremost Girton College for the three-year Research Fellowship, without which this book would not have been written. Together with the Department of History and Philosophy of Science (University of Cambridge), Girton College has been a stimulating cultural environment for the presentation and discussion of my ideas.

Many philosophers, through their writings and discussions, have influenced my views. Steven French offered most valuable comments on my Ph.D. thesis that were crucial for working it up into book form. The several discussions we had about the exclusion principle and related issues have greatly influenced some of the ideas put forward in this book. There is another person who in the past three years has played an important maieutic role in refining my views, and he is Peter Lipton. His insightful comments on earlier versions of some chapters of this book have been most helpful in clarifying my exposition and suggesting possible ways of developing my arguments. John Norton's articles on demonstrative induction originally inspired my Ph.D. thesis: I thank him for illuminating comments on a paper of mine, 'What demonstrative induction can do against the threat of underdetermination: Bohr, Heisenberg, and Pauli on spectroscopic anomalies (1921–24)', *Synthese* **140** (2004), 243–77 (Copyright (2004) by Kluwer Academic Publishers. Reprinted with kind permission from Springer Science and Business Media), which – in an adapted and shortened form – features in this book as Section 3.3.1. Section 5.3.2.1 is a shortened version of my article 'Non-defensible middle ground for experimental realism: we are justified to believe in colored quarks' *Philosophy of Science* **71** (2004), 36–60 (Copyright (2004) by the Philosophy of Science Association. Reprinted with kind permission from the Philosophy of Science Association). I owe also a debt to Marina Frasca-Spada, for comments on an earlier version of Chapter 1 and for initially suggesting the reading of Michael Friedman's *The Dynamics of Reason*. The immense pleasure I took in reading Friedman's book prompted me to rethink some of the main points of my Ph.D. thesis in a refreshingly new way. The reading of Gerd Buchdahl's *Metaphysics and the Philosophy of Science* disclosed a new fascinating perspective for me to explore. I owe the greatest intellectual debt to these two books for the link between a Kantian perspective and the exclusion principle that I have endeavoured to investigate.

I am also grateful to Tian Yu Cao for much helpful advice. Special thanks to Mark Sprevak for innumerable helpful discussions on several points covered in this book, and for much needed support and encouragement during the long and laborious process of writing.

I owe a very special thank you to Stephen Adler, not only for illuminating comments and bibliographic references on the theoretical development of quantum chromodynamics, but also for reading the entire manuscript and for detailed, constructive comments on it. I cannot stress enough how many details Stephen Adler has contributed. Without his invaluable and extremely generous help, I never would have succeeded in attaining the modest level of understanding of quantum chromodynamics presented in Chapter 5. Elie Zahar offered thought-provoking comments on Chapters 1 and 2, for which I am particularly grateful. My Girtonian fellow and friend Peter Sparks commented extensively on the entire manuscript, and very generously gave me assistance with the proofreading. I cannot detail the innumerable improvements he brought to the text, and how much I valued his help. I thank the staff of the Godfrey Lowell Cabot Science Library at Harvard for kindly permitting the inter-library loan of Kuhn's videotape on 'The crisis of the old quantum theory: 1922–25'.

Intellectual debts aside, I owe the major debt to my parents. Their constant encouragement, care, and immense love have always sustained me. 'Thanks' is not the word. This book is dedicated to them, with unspeakable gratitude and love.

# Epigraph

When Galileo rolled balls of a weight chosen by himself down an inclined plane ... a light dawned on all those who study nature. They comprehended that reason has insight only into what it itself produces according to its own design; that it must take the lead with principles for its judgements according to constant laws and compel nature to answer its questions, rather than letting nature guide its movements by keeping reason, as it were, in leading-strings; for otherwise accidental observations ... can never connect up into a necessary law, which is yet what reason seeks and requires. Reason, in order to be taught by nature, must approach nature with its principles in one hand ... and, in the other hand, the experiments thought out in accordance with these principles – yet in order to be instructed by nature not like a pupil, who has recited to him whatever the teacher wants to say, but like an appointed judge who compels witnesses to answer the questions he puts to them ... This is how natural science was first brought to the secure course of a science after groping about for so many centuries.

*Immanuel Kant*
*Critique of Pure Reason*, Preface to the second edition, B xiii–xiv

# Introduction

> The history of the exclusion principle is already an old one, but its conclusion has not yet been written.[1]

It is now eighty years since Wolfgang Pauli introduced an 'extremely natural' prescriptive rule, while dealing with some spectroscopic anomalies that beset physicists in the heyday of the old quantum theory. The rule excluded the possibility that any two bound electrons in an atom were in the same dynamic state, identified by a set of four quantum numbers. Hence the name of *Ausschließungsregel* (exclusion rule), or Pauli's *Verbot* (Pauli's veto) as Werner Heisenberg nicknamed it. The far-reaching physical significance of this rule became clear only later.

From spectroscopy to atomic physics, from quantum field theory to high-energy physics, there is hardly another scientific principle that has more far-reaching implications than Pauli's exclusion principle. It is thanks to Pauli's principle that one obtains the electronic configurations underlying the classification of chemical elements in Mendeleev's periodic table as well as atomic spectra. To this same principle we credit the statistical behaviour of any half-integral spin particles (protons, neutrons, among many others) and the stability of matter.[2] Shifting to high-energy physics, it is the exclusion principle that fixes the crucial constraint for

---

[1] Pauli (1946), p. 215
[2] On this result, established by the seminal proof of Dyson and Lenard (1967, 1968) and Dyson (1967), and later by the simplified Lieb–Thirring proof (1975), see Lieb (1991). The theorem, whose original proof was valid only in the non-relativistic domain and in the absence of gravitational interactions, shows that 'if $N$ charged non-relativistic point particles belonging to a finite number of distinct species interact with each other according to Coulomb's law, and if the negatively charged particles of each species satisfy the exclusion principle, then the total energy of the system cannot be less than $(-AN)$ where $A$ is a constant independent of $N$', Dyson (1996), p. 32. Although this constitutes a very important application of the exclusion principle, for reasons of space I have not addressed it in this book, together with many other applications, to focus instead on a historico-philosophical analysis of the principle. In

binding quarks in hadrons, which together with leptons compose our physical world.

This book advances a philosophical analysis of the enduring and far-reaching validity of Pauli's principle. It does not aim to address what a scientific principle is. It addresses instead the following epistemological question: under what conditions are we justified in regarding an empirical and contingent rule as a scientific principle? Since the exclusion principle was born as a phenomenological rule in a period of crisis for the old quantum theory, we need to explain how it could be accredited as an important scientific principle of the new quantum theory (after 1925). And only by exploring the function Pauli's rule played in the quantum mechanics framework can we shed light on its distinctive nomological feature.

A historical investigation will accompany and support the philosophical analysis. In the following chapters I reconstruct the historical evolution of the exclusion principle across three main phases: from the original spectroscopic context (1920–4, Chapter 2), to the building-up of the quantum mechanics framework within which the spin–statistics theorem was proved (1925–40, Chapter 4), to the development of quantum chromodynamics and parastatistics which opened the door to experimental tests of the principle (1960s–90s, Chapter 5). As the historical reconstruction will highlight, Pauli's rule attained the status of a scientific principle in virtue of the *regulative* function it played in the quantum mechanics framework broadly construed. Regulative function as opposed to constitutive function: this distinction has a distinguished philosophical pedigree in Kant and in the neo-Kantian tradition of Ernst Cassirer. I latch my analysis of the exclusion principle onto this tradition, of which I give an overview in Chapter 1. Accordingly, I shall propose a version of 'dynamic Kantianism' that focuses on the regulative rather than on the constitutive function of scientific principles, and as such can be seen as complementing Michael Friedman's dynamic Kantianism,[3] while owing an obvious debt to Gerd Buchdahl's reading of Kant.[4]

The reasons for this choice reside in the nature and peculiar history of the exclusion principle itself. Pauli's rule did not play any constitutive role, in Hans Reichenbach's sense of coordinating a mathematical formalism with the physical part of a scientific theory so as to provide the conditions

---

the end, this book is not meant to be a comprehensive physics monograph on the exclusion principle in all its physical aspects and applications, but rather a monograph on its philosophical status as a scientific principle.

[3] Friedman (2001).    [4] Buchdahl (1969a), (1969b).

of possibility of the theory. In this respect, it was a *sui generis* principle as compared to others such as the light principle[5] and the equivalence principle of special and general relativity.[6] Pauli's principle was introduced as a tentative rule on mainly phenomenological grounds, within the semi-empirical discipline of spectroscopy, and in a context of revolutionary transition characterized by the waning fortunes of the old quantum theory. It was from spectroscopic phenomena, with the help of some theoretical assumptions of the old quantum theory, that Pauli's rule was originally derived. Once suitably reinterpreted in the new quantum theory framework as a prescription to antisymmetrize the wave function (Fermi–Dirac statistics), it could accomplish a Kantian regulative function: it granted 'systematic unity' to the increasing body of quantum mechanical knowledge, and in so doing it made it possible to derive new laws (e.g. the spin–statistics theorem), whose effect was to enlarge in turn the nomological[7] scope of the principle itself. Systematic unity is an open-ended regulative goal of scientific inquiry. Necessity and nomological strength accrued to Pauli's rule in virtue of the regulative function it accomplished.

Under the action of permutation invariance, the exclusion principle in its Fermi–Dirac reformulation turned out to be one among several possible quantum statistical prescriptions. The larger mathematical structure disclosed by permutation invariance paved the way for the experimental validation of the principle, thanks to the development of the rival programmes of parastatistics and quantum chromodynamics in the 1960s, both prompted by some prima facie negative evidence against the exclusion principle in the context of the quark theory. This process of validation was a two-way street. On the one side, it passed through the parastatistics prediction of Pauli-violating states or more precisely through the prediction of hypothetical particles called 'parons', which supposedly violated the principle by a small amount. This led to the first rigorous experimental tests of the principle more than sixty years after its introduction. It passed, on the other side, through the empirical support that quantum chromodynamics, based on the Pauli-obeying 'coloured' quarks, received.

---

[5] 'Light principle' denotes the special relativity principle that fixes the finite value for the velocity of light.
[6] For Reichenbach's notion of coordinating principles and its relation to Friedman's analysis of the constitutive role of these two principles, see Section 1.3.
[7] 'Nomological scope' denotes the scope of applicability of Pauli's principle as a scientific principle.

My version of dynamic Kantianism owes much to Ernst Cassirer, not only in the priority given to the regulative over the constitutive function, but also in the reappraisal of scientific rationality it delivers. This book is in the end an essay on why we are justified in believing in the exclusion principle and on the rationale for it. This rationale is rooted in the history of the principle. What then needs to be explained in the first place is how the exclusion principle became a live option in the period of revolutionary transition around 1924, when the old quantum theory was proving increasingly inadequate and a new framework had still to be developed. Second, what is the rationale for the enduring validity of the exclusion principle, in the face of some negative evidence. Chapters 3 and 5 are respectively dedicated to these two aspects.

In Chapter 3, following on the historical reconstruction of Chapter 2, I shall argue that the exclusion rule and the associated concept of the electron's spin became a live option for physicists around 1925 because of the piecemeal process of transformation of the old quantum theory into the new quantum theory. Against Kuhn's much celebrated incommensurability thesis, I shall defend the prospective intelligibility of the revolutionary transition around 1925. I shall focus on Kuhn's later writings, where incommensurability is redefined as untranslatability between scientific lexicons, and assess Kuhn's argument, its hidden lemmas and credentials. I shall conclude that the electron's spin and Pauli's exclusion rule were live options for physicists still working with the old quantum theory because they were derived from anomalous phenomena with the help of some theoretical assumptions of the old quantum theory itself, no matter how shaky the grounds these were on already. Newton's method of deduction from phenomena will then provide the methodological framework for defending the prospective intelligibility of this revolutionary shift.

Having reconstructed in Chapter 4 how Pauli's rule became an important principle of the new quantum theory, in Chapter 5 I shall concentrate on the rationale for retaining Pauli's principle in the face of recalcitrant evidence. Against the Duhem–Quine thesis, and more precisely against the threat of underdetermination of theory by evidence that Quine's epistemological holism seems to deliver, I shall argue for the rationale of retaining Pauli's principle in the quark theory, despite some negative evidence. This rationale resides in the regulative function of the principle within a system of knowledge where phenomena are the starting point and final point. In the end, the enduring nomological validity of the principle rests on experimental tests and on the empirical support that accrues to it through various channels.

This view retains a Kantian 'internal' element in distancing itself from traditional scientific realism as well as from antirealism. It does not involve a top-down realist approach (from theory to reality), but rather a bottom-up approach that starts from phenomena to deduce rules, which can then be accredited in virtue of their regulative function within a body of knowledge, and recursively strengthened via other phenomena and larger mathematical structures in a bootstrapping process. It is via this process that epistemic access to scientific entities is disclosed. In the Conclusion I foreshadow this complex issue, which goes far beyond the scope and purpose of this monograph, and I indicate possible lines of future research.

I have written this book not only for my fellow philosophers of science, but also for historians of science, and for physicists. This book occupies a border territory. Although the aim is to advance a philosophical analysis of the role and function of the exclusion principle, it also raises some points about Kuhnian incommensurability and the rationality of inter-framework shifts that apply to physics in a way that does not differ significantly from the way they may apply to social sciences. This book originates from talking to physicist friends and looking with never-ending philosophical curiosity at their work. I hope the book will also serve a different kind of audience, namely graduate students and advanced undergraduates both in philosophy and in physics. I have always thought that philosophy of science should give more of a role to history of science as well as to science itself: this book is an attempt to do philosophy of science with an eye towards both history and physics. Physics courses usually do not have time for the philosophical questions students may find themselves asking as they progress in learning physics. This book is written also for them.

Given the variety of interwoven themes, each audience will probably find parts of this book difficult to read. Summaries are placed at the beginning of each chapter to allow readers approaching the book from different perspectives to follow the overall line of argument without necessarily going through all sections. There is much philosophical, historical, and physical literature that this book assumes. Even so, there are important aspects that, for reasons of space, I could not address.[8] I have tried to write a book that could effectively bring together philosophy, history,

---

[8] For instance, I shall not address metaphysical issues concerning Pauli's principle, namely the issue of identity and individuality of indistinguishable particles obeying the exclusion principle, and the related Leibniz's principle of identity of indiscernibles, which is at the centre of an ongoing debate in philosophy of physics. See French and Redhead (1988), Redhead and Teller (1992), Massimi (2001).

and physics, without reducing interdisciplinarity to a hollow sloganeering. The reader will find some chapters eminently historical, others purely philosophical; in both cases I have endeavoured to keep the discussion at a physically informed level, without going into painstaking physical details. I sense that the final result still cannot do complete justice to the complexity of the topic in all its different aspects. It remains an attempt to venture into a borderland where different fields merge and important philosophical questions may be raised. The challenge was definitely worth taking up. And if others are attracted to venture into the same borderland to correct my mistakes, the challenge can be considered won.

# 1
# The exclusion principle: a philosophical overview

This chapter sets the scene for the philosophical analysis of the exclusion principle that I shall carry out in this book. What is the role and function of a scientific principle? Whence does it derive its accreditation and nomological strength? In the philosophical literature on scientific principles, different answers have been given to these questions, from Poincaré's conventionalism to Reichenbach's analysis of coordinating principles. More recently, Michael Friedman has latched onto the latter tradition to defend 'relativized a priori principles' as principles that are subject to revision during scientific revolutions, but at the same time maintain a *constitutively* a priori role within a theoretical framework. This is germane to a reinterpretation of Kant's notion of 'a priori', whose purpose is to make a Kantian approach to scientific principles compatible with scientific revolutions and modern scientific developments; whence a resultant 'dynamic Kantianism'. I shall endorse a suitable version of dynamic Kantianism to investigate the nature of the exclusion principle as playing a 'regulative' rather than a 'constitutive' function. The regulative/constitutive distinction has a distinguished philosophical pedigree in Kant and in the neo-Kantian tradition of Ernst Cassirer, as I shall spell out in Section 1.4.

## 1.1 Introduction

In a letter to Alfred Landé on 24 November 1924, Wolfgang Pauli announced an 'extremely natural prescriptive rule' that could shed light on some puzzling spectroscopic phenomena he had dealt with in the past three years. The foundation of the rule remained an open question.

Nevertheless, a few months later, in January 1925, Pauli announced it as if it were a commandment of nature:

> In an atom there cannot be two or more equivalent electrons for which the values of all four quantum numbers coincide. If an electron exists in an atom for which all of these numbers have definite values, then this state is occupied.[1]

The rule excluded the possibility for any two bound electrons in an atom to be in the same dynamic state, characterized by a set of four quantum numbers.[2] Hence the name *Ausschließungsregel*: exclusion rule. One year later, in 1926, independently of each other, Fermi and Dirac gave a more precise mathematical formulation to the rule by noticing that restriction to antisymmetric state functions implied it. The rule was accordingly reformulated as prescribing the mathematical nature of quantum states allowed for electrons: it excluded all classes of mathematically possible solutions of the wave equation for any two electrons different from the antisymmetric one. The resultant Fermi–Dirac statistics allowed a system of indistinguishable particles obeying Pauli's principle ('fermions') to be only in antisymmetric states.

When in 1940 Pauli proved the spin–statistics theorem, it became clear that not only electrons, but in fact *any* half-integral spin particle obeyed the Fermi–Dirac statistics, and hence the exclusion principle. The impact of this result for subsequent scientific developments is striking: as we shall see in Chapter 5, when quarks were introduced in the 1960s, they were taken as particles obeying the exclusion principle, given their half-integral spin and the spin–statistics connection established by Pauli's theorem. The discovery of some prima facie negative evidence against quarks obeying Pauli's principle gave rise to two rival research programmes: the parastatistics programme that revoked the strict validity of the exclusion principle for quarks; and quantum chromodynamics that on the contrary reconciled the negative evidence by introducing a further degree of freedom ('colour') for quarks. It was precisely the development of these two rival research programmes that, in different ways, strengthened the nomological validity of Pauli's principle.

---

[1] Pauli (1925b), p. 776.
[2] The four quantum numbers in the modern notation are $n, l, m, s$. The principal quantum number $n$ defines the energy level of the electron; the azimuthal quantum number $l$ measures the orbital angular momentum; the magnetic quantum number $m$ represents the possible orientations of the electron's orbit with respect to a magnetic field; and the fourth quantum number is the spin $s$. Notice that these are not the quantum numbers Pauli used in his article, as we will see in detail in Chapter 2.

Why and how could Pauli's rule – tentatively introduced to deal with some puzzling spectroscopic phenomena – become a building-block of physics, whose validity sweeps across nuclear and atomic physics, from condensed matter physics to quantum chromodynamics? Only by exploring the function Pauli's rule accomplished within the quantum mechanics framework can we shed light on its distinctive nomological features. To this purpose, a historical analysis is required. I shall focus on three main phases in the history of the principle: (i) its origin in the context of spectroscopy around 1920–4 (Chapter 2); (ii) its embedding into the quantum mechanics framework, from which the spin–statistics theorem was later derived (1925–40) as I shall reconstruct in Chapter 4; and (iii) the development of quantum chromodynamics and of the rival parastatistics programme in the 1960s, which paved the way for recent experimental tests of the principle (1960s–90s), to which Chapter 5 is dedicated. Along this historical path, we will find records of now forgotten physical concepts (e.g. Sommerfeld's inner quantum number); discarded models (from the atomic core model to the semi-classical spinning electron model); and novel, undreamt-of scientific entities (coloured quarks). The history of the exclusion principle cuts across the ups-and-downs of twentieth century physics, across its great achievements as well as some of its once popular but now dismissed ideas.

Before any analysis can be undertaken, an overview of the philosophical literature on scientific principles is required. We have to go back to the beginning of last century, when the breakthroughs of relativity theory and quantum mechanics stimulated and prompted philosophical investigations. Henri Poincaré and Hans Reichenbach advanced significant analyses of scientific principles, which only the repeated attacks and the subsequent demise of conventionalism and logical positivism made philosophers forget about. Michael Friedman has very recently offered a long overdue reappraisal of this literature, whose philosophical significance is still so relevant. The bulk of this chapter is dedicated to this literature. It will provide me with a foil to clarify how my analysis of the exclusion principle latches onto a time-honoured philosophical tradition.

## 1.2 From Poincaré's conventionalism to Popper and Lakatos on the nature of the exclusion principle

Henry Poincaré's investigation of scientific principles was prompted by the role that the so-called Physics of the Principles played in his 'structural realist' view. Poincaré believed that the objectivity of science resides in some (almost) permanent relations among the mathematical structures of subsequent scientific theories. For instance, the structural continuity between Fresnel's ether theory – no matter how ontologically false the hypothesis of ether was – and Maxwell's electromagnetic theory was warranted by some fundamental physical principles such as the principle of conservation of energy and the principle of least action.[3] Poincaré's structural realism is hardly conceivable without the Physics of the Principles. And this urged an analysis of scientific principles as bridging the gap between subsequent theories.

The distinctive features Poincaré attributed to scientific principles were certainty and permanence through scientific developments. Poincaré explained these two features in terms of the usefulness of the principles, more precisely in terms of their being useful *conventions*:

> If these postulates possess a generality and a certainty which are lacking to the experimental verities whence they are drawn, this is because they reduce in the last analysis to a mere convention which we have the right to make, because we are certain beforehand that no experiment can ever contradict it.[4]

In contrast with the more radical conventionalist approach of LeRoy, who claimed that science consists of wholly arbitrary conventions,[5] Poincaré insisted that conventions are not at our arbitrary caprice, but are adopted because some experiments have shown that they would be useful and their contraries would not generally succeed. A convention is chosen whenever scientists deal with experimental situations that apparently defy laws of nature. Suppose, for instance,[6] that astronomers discover that stars (call them A) do not exactly obey Newton's law of gravitation (call it B). As a result, they can decide to question *either* that gravitation varies exactly as the inverse of the square of the distance *or* that gravitation is the only force acting on stars. But, Poincaré claimed, the tension between A and B is resolved by introducing an intermediary C (the very notion of gravitation)

---

[3] 'We know nothing as to what the ether is, how its molecules are disposed, whether they attract or repel each other; but we know that this medium transmits at the same time the optical perturbations and the electric perturbations; we know that this transmission must take place *in conformity with general principles of mechanics*, and that suffices us for the establishment of the equations of [Maxwell's] electromagnetic field', Poincaré (1905), English translation (1982) p. 301, emphasis added. See also Poincaré (1902), Chapter XII, Section 3. For the complete list of principles, see the section entitled 'The Physics of the Principles' in Poincaré (1905), English translation (1982) pp. 299–301.
[4] Poincaré (1902), English translation (1982), p. 124.
[5] For a criticism of LeRoy, see Poincaré (1902), Part III, Chapter 10.
[6] See Poincaré (1905), English translation (1982), pp. 334–5.

so that the original relation A–B (stars obey Newton's law) splits into two subrelations:

(i) the relation A–C (stars are acted upon only by gravitation) and
(ii) the relation C–B (gravitation obeys Newton's law).

The first is an experimental law subject to test and revision; the second is a principle which is neither true nor false, but just convenient to make rigorous the previous experimental law A–B, which was only approximately true. Even if the principle is drawn from an experimental law, it is not itself an experimental law, but a definition of gravitation, and as such it is not subject to experimental testing.

Thus, Poincaré contended that scientific principles – despite their experimental origin – are unassailable by experiment because they are mere conventional definitions: experimental tests are either impossible[7] or can be easily accommodated, in the case of a negative result, by putting the blame on some auxiliary hypotheses rather than on the principle itself.[8]

Poincaré's conventionalism offered an intriguing analysis of the nature of scientific principles. But the controversial claim that scientific principles are experimentally invulnerable soon became an easy target. While conventionalists aimed to ground the system of knowledge upon ultimately untestable conventions,[9] for Karl Popper the aim of science was not to discover absolute certainty, but on the contrary to discover theories that could be subject to severe tests. In Popper's view, the gold standard of scientific theories is falsifiability: only through it does science progress.[10]

According to this alternative view, scientific principles are then vetoes more for the technician (e.g. 'there are no perpetual motion machines of the second kind') than for the scientist.[11] The imperative of falsifying scientific theories has as a counterpart the imperative of never making use of what Popper called 'conventionalist stratagems'[12] to save a scientific theory at any cost. Among these stratagems, a central position is occupied by auxiliary hypotheses introduced to accommodate negative evidence and to shelter the theory from falsification. However, Popper conceded that not all auxiliary hypotheses are conventionalist stratagems. Some of them are acceptable, even

---

[7] This is for instance the case of the principle of inertia. See Poincaré (1902), English translation (1982), pp. 96–7.
[8] Curie's calorimetric experiment on radium, which seemed to violate the principle of conservation of energy, could be accommodated by passing the buck back to the chemical nature of radium and leaving intact the principle. See Poincaré (1905), English translation (1982), p. 318.
[9] See Popper (1934), English translation (1968), p. 80.   [10] Popper (1972), p. 361.
[11] *Ibid.*   [12] See Popper (1934), English translation (1968), p. 82.

'eminently acceptable'. In Popper's view, the eminently acceptable auxiliary hypotheses are those whose introduction does not diminish but rather increases the degree of falsifiability of the theory. The degree of falsifiability of a theory is given by the class of its potential falsifiers, i.e. the class of those statements which are forbidden by the theory so that their occurrence would falsify it. More precisely, let T* be a theory obtained from a previous theory T with the introduction of an auxiliary hypothesis H. If T*'s class of potential falsifiers includes T's class of potential falsifiers as a proper subclass, then H has increased the degree of falsifiability of T.

As an example of an eminently acceptable auxiliary hypothesis in the sense just clarified, Popper mentioned the exclusion principle:[13] by forbidding the occurrence of two electrons in the same dynamic state, the exclusion principle has increased the degree of falsifiability of Bohr's atomic theory. By contrast, the Lorentz–Fitzgerald contraction hypothesis was a highly unsatisfactory auxiliary hypothesis because it was ad hoc introduced to shelter Fresnel's ether theory from the negative evidence of the Michelson–Morley experiment.

Despite the declared polemic intent against conventionalism, Popper's methodological falsificationism was indeed close to conventionalism – as Imre Lakatos pointed out – insofar as:

> Popper agrees with the conventionalists that theories and factual propositions can always be harmonized with the help of auxiliary hypotheses: he agrees that the problem is how to demarcate between scientific and pseudoscientific adjustments ... According to Popper, saving a theory with the help of auxiliary hypotheses which satisfy certain well-defined conditions represents scientific progress; but saving a theory with the help of auxiliary hypotheses which do not, represents degeneration.[14]

On Lakatos' methodology of scientific research programmes (MSRP), the introduction of auxiliary hypotheses can make a series of theories (theoretically and empirically) *progressive* or *degenerating*, where a series of theories is progressive if:

(a) each new theory has excess empirical content over its predecessors, i.e. it predicts novel facts (theoretically progressive);
(b) the excess empirical content is in part confirmed, i.e. some new facts are discovered (empirically progressive);

and it is degenerating if it does not meet (a)–(b).

---

[13] *Ibid.*, p. 83.  [14] Lakatos (1978), p. 33.

Accordingly, fruitfulness rather than falsifiability is Lakatos' requirement for an auxiliary hypothesis to be eminently acceptable. If it satisfies this requirement, the hypothesis is positively introduced in what Lakatos called the protective belt of auxiliary hypotheses sheltering the hard core of the theory. Since the core is irrefutable by methodological decision of its proponents, any eventual piece of negative evidence is dealt with by modifying or replacing auxiliary hypotheses. As an example of an auxiliary hypothesis satisfactorily introduced in the protective belt to yield a progressive shift in the theory, Lakatos too mentioned the exclusion principle. In Lakatos' words, the principle was 'invented' by Pauli to shelter the hard core of Bohr's atomic theory – in its mathematically improved version due to Sommerfeld – from negative spectroscopic evidence. Thanks to this beautiful, original, and empirically successful auxiliary hypothesis, which 'accounted not only for the known gaps [in Sommerfeld's spectroscopic theory] but reshaped the shell theory of the periodic system of elements and anticipated facts then unknown',[15] Bohr's programme was (at least temporarily) safeguarded.

But was the exclusion principle really 'invented' as an auxiliary hypothesis – albeit a very satisfactory one – to shelter Bohr's theory or to increase its degree of falsifiability? There is a historical problem concerning whether Popper's or Lakatos' rational reconstruction of the origin of the exclusion principle is faithful to the real history. Clearly, it is not. As we shall see in Chapter 2, the exclusion principle was *not* introduced to shelter Bohr's theory from negative evidence, nor could it actually do so. In fact, the principle could not be reconciled with Bohr's correspondence principle, and Pauli repeatedly underlined this point as the sign that a new quantum theory was required to ground his rule. Leaving aside this issue for the moment, it suffices here to say that the new philosophical tendencies associated with the names of Popper and Lakatos contributed much to the demise of conventionalism. Not just to that. The second scapegoat was logical positivism, which had also offered an interesting philosophical perspective on scientific principles, to which I now turn.

## 1.3 From Reichenbach's coordinating principles to Friedman's relativized a priori principles

The once-dominant movement of logical positivism was extensively criticized (by Popper, Kuhn, and Quine, to mention just a few names) to the

---
[15] *Ibid.*, p. 67.

extent that the philosophical significance and the original impact of the movement was increasingly underestimated. In recent times, Michael Friedman has been one of the very few voices in the philosophical panorama to offer a reappraisal of logical positivism in the light of the original cultural context in which the movement flourished.[16] This was the context of the Weimar culture, where new philosophical streams grew and developed under the influence of recent scientific breakthroughs. The introduction of non-Euclidean geometries had opened the door to Einstein's general relativity, and to the consequent demise of Newtonian mechanics, after which Kant's analysis of scientific knowledge had been patterned. As was to be expected, this scientific revolution paved the way to a reconsideration of Kant's philosophy itself. In particular, it was Kant's notion of *synthetic a priori judgments*, the distinctive feature of scientific knowledge, which emerged deeply undermined. Non-Euclidean geometries and Einstein's relativity taking over Newtonian mechanics seriously challenged the Kantian a priori character of scientific knowledge. As became evident, no system of scientific knowledge is fixed once and for all, or immune to revision. Philosophers drew their conclusions. Poincaré's conclusion was conventionalism. Hans Reichenbach's conclusion was that the notion of a priori needed to be reconsidered in such a way that a Kantian kernel could be retained without the allure of non-revisability.

In *The Theory of Relativity and A Priori Knowledge*[17] Reichenbach distinguished two meanings in the Kantian notion of a priori: (1) unrevisable, fixed once and for all; (2) *constitutive* of the object of experience.[18] According to Reichenbach, the recent scientific developments had revoked the first meaning, but not the second. This was particularly evident in the case of what Reichenbach called *axioms of coordination*. They are non-empirical (and then non-synthetic) principles that coordinate the mathematical part of a scientific theory with the physical part of the theory. For example, in Newtonian mechanics, Newton's three principles count as axioms of coordination or coordinating principles between the Euclidean space, which is the proper geometrico-mathematical part of the theory, and Newton's gravitational law, which is the physical and experimentally testable part of the theory. In this specific sense, Newton's principles can be regarded as *constitutive* of Newtonian mechanics: they need to be in place together with the mathematical part, for the physical part of Newton's theory to be possible at all. As such, Newton's principles are a

---

[16] Friedman (1999). [17] Reichenbach (1920). [18] See Friedman (1999), p. 61.

## 1.3 From Reichenbach to Friedman

priori in the second sense: they are *constitutively* a priori, not a priori as synonymous with fixed and unrevisable.

With the development of relativity theory, new axioms of coordination have replaced Newton's as the *constitutively a priori* principles of the new framework: namely, the light principle of special relativity and the equivalence principle of general relativity. The former coordinates the four-dimensional Minkowski space-time with Maxwell's equations in the special relativity framework. The latter coordinates the Riemannian space-time manifolds with Einstein's equations for the gravitational field in general relativity.

Along the lines of Reichenbach, Michael Friedman has recently defended *relativized a priori* principles as, on the one hand vindicating the Kantian idea of *constitutively a priori* elements of a scientific framework, and on the other as *relative* to it and hence revisable during a scientific revolution; hence a resultant form of 'dynamic Kantianism'.[19] 'Dynamic Kantianism' is a fascinating attempt to reconsider and adapt Kant's transcendental philosophy to modern science, by disentangling it from Newtonian mechanics and Euclidean geometry, after which it was originally patterned. But how to reconcile the fact that, as *constitutively a priori* elements of the framework, these principles cannot be subject to experimental testing, while as *relativized* they can be modified and replaced as soon as significantly new evidence is discovered?

Relativized a priori principles do have empirical content. And yet they cannot be empirically true or false, confirmed or disconfirmed as empirical laws are: i.e. they cannot be confirmed or disconfirmed within the same framework wherein they play a constitutive function. This view poses the problem of explaining how a relativized a priori principle can emerge in the first place as constitutive of a framework, and how it can later be dismissed during a scientific revolution.

Friedman considers the cases of the Michelson–Morley experiment for the light principle of special relativity, and the experiment of von Eötvös for the principle of equivalence in general relativity. The Michelson–Morley experiment gave negative evidence for the motion of light through ether; but it cannot be taken as an experiment confirming

---

[19] As Friedman (2001) points out, there are quite fundamental differences between this 'dynamic Kantianism' and Kant's original view. Most notably, Kant's conception of pure intuition, and the schematism of the pure concepts of the understanding implies that the understanding can a priori impose a *unique* set of constitutive principles on experience. For an analysis of Kant's philosophy and its relationship to natural sciences, see also Friedman (1992a).

Einstein's light principle. Rather, Friedman argues, it is more accurate to say that 'Einstein uses his light principle *empirically to define* a fundamentally new notion of simultaneity and, as a consequence, fundamentally new metrical structures for both space and time.'[20] It is at this stage that an element of 'decision' comes in to 'elevate' the empirical findings of the Michelson–Morley experiment to the status of a constitutive principle underpinning a new spatio-temporal framework. Similarly for the principle of equivalence. The fact that gravitational and inertial masses are the same (as von Eötvös' experiment showed) was well-known since the times of Newton. Yet, only with Einstein was this already accepted empirical fact 'elevated' to the status of a new constitutive principle for the framework of general relativity, within which Einstein's field equations could be tested and confirmed by the evidence about the anomalous perihelion of Mercury. Moreover, it is this same evidence that indirectly gave the verdict also to the light principle. As Friedman points out, the principle of equivalence and the Riemannian theory of manifolds provided the conditions of possibility not only for general relativity, but also for a rival formulation of the Newtonian theory of gravitation (proposed by Élie Cartan in 1923–4), so that two empirically equivalent theories were faced off against the background of a common constitutive framework. And if the perihelion of Mercury confirmed general relativity at all, this is because it confirmed what distinguishes general relativity from the alternative formulation of Newtonian theory: namely, the (infinitesimally) Minkowskian metrical structure derived from special relativity. So, a non-empirical constitutive principle such as the light principle of special relativity, entirely beyond the reach of standard empirical testing at one stage of scientific progress, can be subject to precisely such testing at a later stage, as is to be expected on a truly dynamical conception of the a priori.[21]

This dynamical conception of a priori principles is all the more relevant to scientific rationality. If the constitutive a priori principles are in the end relative to a certain framework and subject to change during scientific revolutions, is there any rationality in scientific developments? We should here distinguish between a retrospective and a prospective notion of scientific rationality. *Retrospective* rationality can be argued for by looking at the way the new framework embeds the old one as an approximate limiting case. So Euclidean geometry can be regarded as a limiting case of Riemannian manifolds, and Newtonian space-time can be retrieved from special relativity if the velocity of light goes to infinity. *Prospective*

---

[20] Friedman (2001), p. 88.    [21] *Ibid.*, p. 92.

## 1.3 From Reichenbach to Friedman

rationality is more difficult to grasp. How can it be rational for practitioners of an earlier framework to move to a new one with a possibly different mathematical formalism and brand new constitutive principles that may well appear unintelligible and meaningless to them? Here there can hardly be any appeal to experimental evidence. If the perihelion of Mercury confirmed the validity of Einstein's field equations at all, this is only because the new framework of Riemannian manifolds and the principle of equivalence were already a live option for physicists. But how could they become a live option in the first place? Here is where Kuhnian incommensurability creeps into the picture: scientific revolutions resemble more conversion-like experiences than a rational process.

Friedman's answer is meant to mitigate the Kuhnian picture. And, interestingly enough, this answer relies precisely on his analysis of relativized a priori principles as rooted in well-known empirical facts of the earlier framework that at some point were elevated to the rank of principles constitutive of a new framework. It is worth quoting Friedman at some length on this point:

> Despite the fact that practitioners of the new framework indeed speak a language incommensurable or non-intertranslatable with the old, they are nonetheless in a position rationally to appeal to practitioners of the older framework, and to do this, moreover, using empirical and conceptual resources that are already available at precisely this earlier stage ... Einstein here took an already well-established empirical fact (the empirical indistinguishability of different inertial frames by optical and electrodynamical means) and 'elevated' it to the status of a convention or coordinating principle. What he saw, which no one did before, was that this already well-established empirical fact can indeed provide the basis for a radically new coordination of spatial and temporal structure ... He thereby put practitioners of the earlier physics in a position both to understand his introduction of a radically new coordination and, with a little good will, to appreciate it and to accept it as a genuine alternative ... To gain acceptance of the new framework merely as a rational (real) possibility – as a reasonable and responsible live option – is already more than half the battle.[22]

Thus, Friedman entrusts the prospective rationality of science to relativized a priori principles, namely to their being well-known empirical facts that at some point were 'elevated' to the status of constitutive principles of a new framework. Friedman's account of prospective inter-framework rationality is very attractive and insightful. It gives Kuhn its due in the *relativized* conception of a priori principles, while at the same time it does justice to the Enlightenment ideal of the rationality of science as cutting

---

[22] *Ibid.*, p. 101–3.

across revolutionary changes. I am sympathetic with Friedman's dynamic Kantianism, but I want to point out a few problematic aspects, before advocating an alternative version of dynamic Kantianism in later sections.

One of the pillars of Michael Friedman's dynamic Kantianism is a criticism of Quine's epistemological holism. Quine has vividly portrayed epistemological holism in the following famous passage:

> The totality of our so-called knowledge or beliefs, from the most casual matters of geography and history to the profoundest laws of atomic physics or even of pure mathematics and logic, is a man-made fabric which impinges on experience only along the edges ... A conflict with experience at the periphery occasions readjustments in the interior of the field ... But the total field is so underdetermined by its boundary conditions, experience, that there is much latitude of choice as to what statement to re-evaluate in the light of any single contrary experience ... Furthermore, it becomes folly to seek a boundary between synthetic statements, which hold contingently on experience, and analytic statements, which hold come what may. Any statement can be held true come what may, if we make drastic enough adjustments elsewhere in the system ... Conversely, by the same token, no statement is immune to revision.[23]

Quine's epistemological holism challenges the time-honoured analytic/synthetic distinction. While claiming that nothing is forever immune to revision (not even the laws of mathematics and logic), Quine nonetheless allowed beliefs at the centre of the web to be less promptly revisable simply because more distant from experience. Revision is in the end only a matter of 'varying distances from a sensory periphery', and this reflects nothing more than 'the relative likelihood, in practice, of our choosing one statement rather than another for revision in the event of recalcitrant experience'.[24]

Friedman has questioned Quine's claim that a scientific theory faces the 'tribunal of experience' as a whole, by stressing instead the asymmetry between the mathematical part and the proper physical part of a scientific theory. Since the former, in conjunction with constitutive a priori principles, provides the conditions of possibility for the physical part, it follows that the two cannot be happily viewed as a symmetrically functioning conjunction that faces the 'tribunal of experience' as a whole. Nor, on the other hand, can constitutive a priori principles be reduced to central well-accredited elements of a Quinean web of knowledge that some sort of scientific conservatism would prevent us from revising. For instance,

---

[23] Quine (1951). Reprinted in Quine (1953), pp. 42–3.   [24] *Ibid.*, p. 43.

when Newton formulated his theory of gravitation, the mathematics of the calculus was not yet universally accepted, nor were Newton's principles any better accredited at the time. Thus, Friedman concludes, the relativized and dynamical conception of the a priori developed by the logical positivists can describe conceptual revolutions far better than Quinean holism does.[25]

The following dilemma arises. *Either*, following Quine, any part of a scientific theory (including scientific principles) is always open to revision, whereby revision is only a matter of degree or 'varying distances' from evidence, which poses the problem of explaining how revolutionarily new mathematical formalisms and scientific principles can ever arise. *Or*, with Friedman, we can make room for scientific revolutions to occur but at the cost of considering scientific principles as constitutive a priori and hence not subject to experimental testing and revision within the framework wherein they accomplish a constitutive role.

The notion of 'constitutive a priori' seems close to analyticity. Only by echoing the venerable albeit, after Quine, generally dismissed analytic/synthetic dichotomy dear to logical positivists can Friedman argue for the asymmetry between the mathematical part and the physical part of a scientific theory and attribute a special status to principles, whose nomological strength and accreditation do not seem to depend on their being empirically well-supported, but rather on their playing a constitutive function within that framework. And as there is an element of decision for their being 'elevated' to such a status, similarly there must be an element of decision for their being 'removed' from it during a scientific revolution. Empirical evidence cannot ultimately give the verdict to a constitutive a priori principle, or decide its fortune. What empirical evidence does is simply to guarantee that they have empirical content and originate from well-known empirical facts. But once a new framework with its constitutive a priori principles has been put forward, the nomological strength and the accreditation of the principles seem to be independent of empirical evidence.

Yet, this philosophical position seems in contrast with scientific practice where no clear-cut distinction exists between parts of a scientific theory that are in principle revisable and those that in principle are not. Surely, it is more appropriate to say with Quine that experimental revision is only a matter of degree, rather than a principled distinction. Furthermore, I think we should give more of a role to empirical evidence in the process of

---

[25] Friedman (2000a), pp. 376–7.

20          *1 The exclusion principle: a philosophical overview*

validation and accreditation of scientific principles. Their nomological strength does not depend on their constitutive function any more than on the degree of empirical support that a Quinean-like scientific web may accrue to them, as I shall argue in Chapter 5 in the specific case of the validation of the exclusion principle.

But going back to Friedman's dynamic Kantianism, there is a second problematic aspect related to this first one: namely, how a well-known empirical fact may ever attain the status of a constitutive a priori principle for a new framework. Friedman's solution in terms of 'elevating' the empirical fact to such a status seems to give a conventionalist gloss, as if the nomological status of constitutive a priori principles depended in the end on a conventional decision of scientists to promote an empirical fact to such a rank. And this gloss is unfortunate for two different reasons. Saying that constitutive a priori principles are relative to a framework (as opposed to being fixed and unrevisable for all time) does not necessarily imply that they are *conventional*. In particular, conventionalism seems at odds with the *constitutive* nature of these principles: they cannot be *conventionally* constitutive if we want to stick to a genuinely Kantian perspective, albeit a dynamic one. Furthermore, this conventionalist gloss risks putting the burden of scientific revolution mainly on the genius of a single scientist, for instance Einstein, whose acumen envisioned the possibility of using a well-established fact (the equality of gravitational and inertial mass ascertained by von Eötvös) as the basis for a completely new theory of gravitation such as general relativity. But scientific revolutions are hardly ever the product of a single man's genius or visionary insight. It was not Copernicus that made the Copernican revolution, but rather the following generations of people like Brahe, Kepler, Galileo, and Newton who built on Copernicus' tentative steps.[26] Nor was it Planck's quantum postulate that made the quantum mechanics revolution, but rather Einstein's use of quanta in his 1905 paper, which opened up a new era for Bohr's atomic theory in 1913, Sommerfeld's re-elaboration of it in 1916, and so forth.[27]

This leads me to a third problematic aspect of Friedman's account. In Friedman's view, prospective rationality is secured at the cost of understating the *revolutionary* nature of scientific principles. Constitutive a priori principles are presented as the joints of two otherwise

---

[26] See Kuhn (1957).
[27] For the history of the origins of quantum theory as related to the black-body problem, see Kuhn (1978).

incommensurable frameworks, thus granting continuity and prospective rationality, as if scientific revolutions happened in their surroundings – so to speak – without involving the principles themselves. But I think instead that scientific principles are the crucial interstitial elements of inter-framework shifts where scientific revolutions primarily take place. For a new framework to arise, a revolution must take place in the scientific principles of the preceding framework. The result is brand new principles that set up the new scene of inquiry. Of course, the origins of the new principles can be looked for in the earlier framework, in the preceding scene of inquiry with its puzzles to be solved, and increasingly inadequate theoretical concepts. It is one thing to claim that revolutionarily new principles have a long incubation period whose deep roots lie in the preceding framework. But it is another thing to say that the new principles are not, strictly speaking, *revolutionarily* new: they are instead old and well-known empirical facts promoted to the rank of constitutive principles of the new framework.

In this monograph on the exclusion principle, I shall endorse an alternative version of dynamic Kantianism, one that does not hinge on constitutive a priori principles. This alternative version of dynamic Kantianism is meant to complement, not to contradict, Friedman's. It is the nature of the exclusion principle that lends itself to this alternative version. By contrast with the light principle or the equivalence principle that emerged from the beginning as cornerstones of special and general relativity, the exclusion principle has more humble origins. It was introduced as a phenomenological rule in the spectroscopy of the 1920s. Its wide-ranging nomological scope did not become evident before the development of Fermi–Dirac statistics and of the spin–statistics theorem in 1940, which remarkably enlarged the scope of the principle from electrons to any half-integral spin particle, opening up unexpected research avenues in the following decades. Before turning to the exclusion principle, let me first clarify the Kantian distinction between constitutive and regulative, which is the kernel of the alternative version of dynamic Kantianism that I shall endorse for my analysis of the exclusion principle.

## 1.4 Constitutive versus regulative

The distinction between constitutive and regulative occupies a central position in Kant's philosophy. In the *Critique of Pure Reason*, this distinction runs parallel to the distinction between the faculty of understanding and the faculty of reason: the former is constitutive of the possibility of experience, the latter has a purely regulative function with respect to experience. In the *Critique of the Power of Judgment*, this distinction is further articulated in a way that assigns priority to the regulative demand, here presented as an expression of the faculty of reflecting judgment. As Michael Friedman has illuminatingly pointed out apropos of this, constitutive principles, due to the faculties of understanding and sensibility, are the necessary conditions of the comprehensibility of the phenomenal world. Regulative principles, on the other hand, due to the faculties of reason and reflecting judgment, can never be fully realized in experience, and present us with ideal ends or goals for seeking the never fully attainable complete science of nature.[28]

The distinction between constitutive and regulative is at the centre of an ongoing debate among Kantian scholars, and not only for exegetical reasons: passages from the first and the third *Critique*, from the *Metaphysical Foundations of Natural Science*, the *Prolegomena*, and the *Opus postumum* lend themselves to different readings that sometimes cannot be easily reconciled. What is mainly at stake in this debate is the conception of lawlikeness in Kant's philosophy of science, hence two divergent interpretations.

On the one side, Michael Friedman has argued for the inseparability of Kant's transcendental philosophy from his commitment to Newtonian mechanics. On this reading, the transcendental principles of the understanding presented in the 'Analytic of Principles' of the *Critique of Pure Reason*, namely what Kant calls the Analogies of Experience (substance, causality, and community),[29] would be tied up with the principles of Newtonian mechanics, the so-called metaphysical principles of pure natural science, namely the conservation of mass, the principle of inertia, and the principle of equality of action and reaction so as to ground the existence and necessity of empirical laws, namely Newton's law of universal gravitation. Kant took as his model Newton's method of deduction from phenomena: the law of universal gravitation was deduced from phenomena, i.e. observed relative motions of the planets as described by Kepler's laws, which were assumed to approximate the true motions (motions with respect to absolute space and time) so that Newton's principles could be applied to them. In this way, the metaphysical principles of

---

[28] See Friedman (2000b), p. 117.
[29] Kant (1781), A182–A218. I use the Guyer and Wood (1997) translation of the *Critique of Pure Reason*, where (A) refers as usual to the pagination of the first edition (1781) of the *Critique* and (B) to the second edition (1787).

## 1.4 Constitutive versus regulative

natural science (i.e. Newton's principles) applied to inductively generalized phenomena allowed the deduction of Newton's law of universal gravitation.[30] Hence, the transcendental principles of pure understanding (Kant's Analogies of Experience) – linked as they are to the metaphysical principles of pure natural science – can be regarded as *constitutive* with respect to experience because they make possible an empirical causal law (the law of universal gravitation), which *qua* empirical is based on inductively generalized phenomena, while *qua* law must have an a priori grounding, namely the one warranted by the transcendental principles. Vice versa, empirical regularities attain nomological status in virtue of their being nested in the metaphysical principles of pure natural science, and then in their transcendental counterparts.

Friedman's intriguing and textually well-supported interpretation has however been accused of 'tightening the noose' for Kant's transcendental philosophy, by making its fortunes dependent on those of Newtonian mechanics. What relevance could Kant's philosophy still have for twenty-first century physics, in the light of the scientific revolutions of relativity theory and quantum mechanics? None of the principles that Kant considered to be synthetic a priori are in fact even correct. To disentangle the enduring significance of Kant's philosophy from the fortunes of Newtonian mechanics, Gerd Buchdahl has defended the 'looseness of fit' between Kant's theory of Newtonian science and the 'Transcendental Analytic' of the *Critique of Pure Reason*, between the empirical laws of physics and Kant's Analogies of Experience. Without entering into details, suffice here to say that a consequence of the 'looseness of fit' interpretive line is a reappraisal of the regulative principles of the faculty of reason over the constitutive principles of the faculty of understanding.

On Buchdahl's view,[31] the lawlikeness of Newtonian mechanics is a function of the architectonic of reason. It is the faculty of reason that would project an order on nature by unifying and systematizing empirical regularities under a hierarchical system of higher-level – albeit still empirical – laws. The theoretical system resulting from this regulative activity of reason gives, in turn, some kind of necessity to laws: the necessity of laws of nature is purely 'injected', it is an expression of the systematizing activity of reason, a function of their role and place in the theoretical system. Hence their double status of being both contingent – since established on an empirical basis – and at the same time necessary, in the sense just clarified.

---

[30] See Friedman (1989), (1992b), (1994).   [31] Buchdahl (1969a), (1969b), (1974).

The necessity of empirical laws would then not be grounded on the transcendental principles of the faculty of understanding as *constitutive*; it would rather be grounded on their specific role and place in a system of knowledge which satisfies a *regulative* demand of reason.

Thus, the constitutive/regulative dichotomy marks a gulf among Kantian scholars as to whether it is the faculty of understanding with its constitutive principles, or rather the faculty of reason with its regulative principles that is ultimately responsible for the law-governedness of nature. The interpretive line of Friedman[32] stresses the former to provide a priori grounding to empirical laws; Buchdahl's alternative interpretation emphasizes the latter to disentangle Kant's philosophical apparatus from any specific scientific theory. In Friedman's view, the regulative demand of reason always functions for the sake of the constitutive demand of understanding. In Buchdahl's, the regulative function stands on its own. And while Friedman himself has underlined the happy convergence of the constitutive and the regulative aspects as an essential feature of Kant's overall philosophical project, especially evident in the *Opus postumum*,[33] his opponents have accused him of giving a lesser role to the regulative function.[34]

It is not my intention or purpose to enter into exegetical and interpretive analyses of Kant: scholarly and authoritative opinions have already been expressed, to which I could add very little. Rather, my aim is to show how this debate bears upon current discussions in philosophy of science. As we saw in the previous section, the dynamic Kantianism that Michael Friedman has recently defended focuses on constitutive a priori principles. I want to draw attention to a different perspective about scientific principles, one that is still *dynamically* Kantian in considering them as relative and revisable, and yet is not distinctively Reichenbachian in identifying them with constitutive a priori principles 'coordinating' the mathematical part with the proper physical part of a scientific theory. The history of Pauli's exclusion principle lends itself naturally to this alternative perspective, which latches onto Friedman's by highlighting the complementary, *regulative* aspect of a Kantian 'dynamics of reason'. In this respect, my analysis follows Buchdahl's interpretation of Kant, and it has a distinguished philosophical pedigree in the neo-Kantian tradition of Ernst Cassirer, as I shall clarify in Section 1.4.2. But let us first have a closer look at Kant, and in particular at his analysis of the regulative principle of 'systematic unity' or systematicity.

---

[32] Along this line, see also Butts (1991).   [33] See Friedman (1991).
[34] See Allison (1994).

### 1.4.1 Kant on the regulative principle of systematicity

Within the regulative domain of the faculty of reason, the principle of systematicity or 'systematic unity' occupies an important position. Kant addressed the issue of systematicity both in the 'Appendix to the Transcendental Dialectic' in the *Critique of Pure Reason*, and in the 'Introduction' to the *Critique of the Power of Judgment*. In the Appendix, systematic unity is presented as an expression of the faculty of reason in its 'hypothetical' as opposed to 'apodictic' use. Reason is the faculty of deriving the particular from the universal: if the universal is in itself certain and given, so that the particular can be derived from it, reason is said to be used in an 'apodictic' way. If, on the other hand, the universal is not given but only assumed problematically as a mere idea, while the particulars are given to be subsumed under it, reason is said to be used 'hypothetically'. Systematic unity is then presented as a regulative idea 'bringing unity into particular cognitions' and hence transforming a 'contingent aggregate' into a 'system interconnected in accordance with necessary laws'.[35] As such, systematic unity is only a 'projected unity', rather than an objective feature of reality.[36]

In the rest of the Appendix, Kant spelled out systematicity in terms of three logical principles: (1) the principle of 'sameness of kind' or *homogeneity*, which enjoins us to gather particulars under higher genera; (2) the principle of variety or *specification*, which is guided by the opposite tendency to unfold the variety of species gathered under the same genus; (3) the principle of affinity of all concepts or *continuity* of forms, 'which offers a continuous transition from every species to every other through a graduated increase of varieties'.[37] The principle of continuity of forms underpins both the principle of homogeneity and specification because only if there is a genuine *continuum formarum* can we rearrange the classification of empirical concepts under either higher-order – that is more abstract – concepts (genera) or lower-order – that is more specific – concepts (species) without any leap in these transitions.[38] These three logical principles as different aspects through which the goal of

---

[35] Kant (1781). English translation (1997) A645/B673, A647/B675.
[36] Although systematicity is not an objective feature of reality, as Guyer (2003) pp. 9–14 has noted, it is however an intrinsic feature of Kant's regulative principles to presuppose that the attainment of a goal is at least conceivable (although it may be impossible in practice). Hence we must believe that there is an actual ground of the objective possibility of the systematic unity of our concepts. For this reason, systematicity is presented not merely as a logical principle but as a transcendental principle, although no transcendental deduction – analogous to the one for the categories of the faculty of understanding – is available for it.
[37] Kant (1781). English translation (1997), A658/B686.   [38] *Ibid.*, A659/B687.

systematization is articulated can themselves never be fully attained, and yet remain as ideals to which we must strive. In fact, Kant says, in nature we find only discrete species; and yet we need to assume these principles and seek under their presumption to maximize unity (according to the principle of homogeneity) among our empirical concepts as well as variety (according to the principle of specification). These two opposite tendencies are only different aspects of reason's interest in approximating the open-ended goal of systematicity.

In the *Critique of the Power of Judgment*, systematic unity is no longer assigned to the faculty of reason, but to the faculty of judgment. This is the faculty of thinking the particular as contained under the universal, where in the 'reflecting judgment' (as opposed to the 'determining judgment') particulars are given and universals must be found under which particulars can be subsumed. The task of reflecting judgment consists then in arranging lower-level empirical concepts and laws into a hierarchical system of higher-level empirical concepts and laws, which are nonetheless still empirical. In this respect, reflecting judgment too is a purely regulative faculty, with its own transcendental principle that postulates the systematic unity of empirical concepts and laws as a regulative idea, so that a system of empirical science is actually possible.[39]

Along lines similar to those of Buchdahl,[40] Paul Guyer has offered a penetrating analysis of systematic unity and of the significance of its reassignment from the faculty of reason to the faculty of reflecting judgment.[41] In particular, Guyer has noticed a shift of emphasis between the first half and the second half of the (unpublished) First Introduction to the *Critique of the Power of Judgment*. In the first half, systematicity is presented as conferring an order and hierarchical organization upon empirical concepts and laws, which are nonetheless discoverable independently of it: it supervenes on empirical laws as the form of their overarching organization. But Kant seems to offer a different picture in the second

---

[39] See on this point Friedman (1991), pp. 74–5.
[40] Apropos the First Introduction to the *Critique of the Power of Judgment*, Buchdahl (1969a), p. 505, had already noted, 'Now it is here that the importance of the concept of system arises, for only such a concept supplies us, Kant claims, with the necessary constraints on the number of possible 'rules' of uniformities, selecting those (as laws) from among infinitely many which can be fitted into such a system. *Per contra*, only those putative uniformities, which can so be fitted will be regarded as laws. It is not just the case that systematization supplies us with the *links between* empirically given laws, but this process is really the very presupposition for any empirical uniformity being regarded as lawlike in any serious sense of the term.'
[41] See Guyer (1990), and (2003) pp. 1–30.

## 1.4 Constitutive versus regulative

half of the Introduction,[42] where systematicity is assigned a more substantial role as the condition of possibility of discovering empirical concepts themselves, and warranting that the variety of natural forms does not exceed our capacity to identify empirical uniformities as lawlike. Thus, Guyer concludes, without systematicity 'we can neither be sure that we can discover laws of nature nor recognise them as laws – that is necessities – if we do recognise them'.[43]

I take this interpretive analysis as a springboard for suggesting that the regulative goal of systematicity is more than an additional desideratum to be glossed over batches of empirical laws, just for the sake of imposing an overall structure and organization on them. Systematicity is instead a necessary condition for lawlikeness itself: it makes it possible for us to find our way through the empirical manifold and to recognize empirical regularities as lawlike. No wonder Philip Kitcher has seen in Kant's analysis of systematic unity the kernel of Kant's philosophy of science, and of its enduring significance for current debates on laws of nature and scientific realism.[44]

Following this interpretive analysis of Kant's principle of systematicity, the upshot of this monograph is to show that it is systematicity that underpins the lawlikeness of the exclusion principle. There is indeed a thin line that runs from Kant's regulative principle of systematicity to the foundations of quantum mechanics, and to the specific role of the exclusion principle in it. This thin line passes through the reflections of a neo-Kantian philosopher, who was also an insightful interpreter of the quantum mechanics revolution. As we shall see in the next section, Ernst Cassirer gave systematicity a new twist in the light of the scientific breakthroughs of twentieth-century physics.

---

[42] I use the Guyer and Matthews (2000) translation of the *Critique of the Power of Judgment* (1790). See First Introduction, Section V, 20: 213.

[43] Guyer (1990), p. 42. This point is further stressed in Guyer (2003), p. 24: 'Here Kant suggests that the problem raised by the possibility of applying the categories to empirical intuition in indeterminately many ways is not that we might not find any empirical concepts for a given manifold at all, nor is it that we might not be able to impose on empirical concepts a systematic order that reason desires for its own sake; rather, it is that in light of such a possibility, we could not see why any particular empirical law, no matter how consistent with the categories on the one hand and the empirical data on the other, should itself be *necessarily true*. Kant assumes, however, that an empirical law must be necessarily true if it is to be a law and that the only ground that we can have for considering any particular empirical law necessarily true is that it occupies a specific position in a *system* of laws.'

[44] Kitcher (1986).

## 1.4.2 Ernst Cassirer and the architectonic of scientific knowledge

In the neo-Kantian tradition, it was Ernst Cassirer who gave a significant reappraisal of the *regulative* over the *constitutive* principles of Kant's philosophy. This is not the place to engage in a discussion of Cassirer's philosophy, as it would lead me far astray from the purpose of this book. But I do want to highlight a few aspects of Cassirer's philosophical view that owe an obvious debt to Kant's discussion of systematicity and, at the same time, are relevant to my philosophical analysis of Pauli's principle.

In the final section of the chapter 'On the problem of induction' in *Substance and Function*, Cassirer noticed that the problem of induction consists in gathering empirical facts and connecting them according to laws of nature, which at first seem to present us with the final and definitive form of experience, whereas in fact they provide us only with a temporary platform for carrying out further analysis as soon as new evidence becomes available.[45] No scientific theory or set of scientific principles can ever be final and definitive: Newton's principles themselves are only the temporarily simplest intellectual 'hypotheses, by which we seek to establish the unity of experience'.[46] They accomplish a *regulative* function. As such, they are not a priori, and enjoy the same empirical status as any other empirical law. What distinguishes them from lower-level empirical laws is simply the unifying and systematizing task they accomplish. This sits squarely within a view of science that seeks systematicity while granting the fully revisable and falsifiable nature of its fundamental principles.

Cassirer reinterprets the 'a priori' in *regulative* terms. The more we strive to isolate the ultimate common element of all possible forms of scientific experience that may be the conditions of possibility of any theory, the more we become aware that at no given stage of knowledge can this goal be perfectly achieved. Yet, it remains a regulative idea, prescribing a fixed direction to the evolution of science. This reinterpretation of the a priori as a regulative idea finds its natural expression in what Cassirer called the 'invariants of experience'.[47] The a priori no longer denotes that which is *prior* to experience in the sense of being the condition of possibility of experience; but rather that which is the ultimate 'invariant' of experience, unattainable at any stage and yet a regulative goal of scientific inquiry. This seminal investigation – carried out in *Substance and Function* –was further articulated and explored in Cassirer's later books, namely in the

---

[45] See Cassirer (1910), English translation (1953), p. 266.  [46] *Ibid.*, p. 268.
[47] *Ibid.*, p. 269.

## 1.4 Constitutive versus regulative

one dedicated to the philosophy of the Enlightenment, and in *Determinism and Indeterminism in Modern Physics*.

In *The Philosophy of the Enlightenment*, Cassirer offered a long overdue reappraisal of the Enlightenment conception of scientific progress and rationality. Against a widespread prejudice, dating back to the verdict of the Romantics, Cassirer stressed instead the distinctive qualitative aspect of the Enlightenment notion of scientific progress, according to which 'one seeks multiplicity in order to be sure of unity ... Variety and diversity of shapes are simply the full unfolding of an essentially homogeneous formative power. When the eighteenth century wants to characterise this power in a single word, it calls it "reason".'[48]

The unifying power of reason was emblematically represented by the work of Isaac Newton. If the seventeenth century with Descartes, Spinoza and Leibniz was dominated by the 'spirit of system' (*esprit de système*), the eighteenth century was on the contrary characterized by its forceful rejection. Against the dogmatic tendency to build up systems, the Enlightenment rediscovered the importance of starting from phenomena. Newton's method of deduction from phenomena (paradigmatically deployed in the *Principia* and in the *Optics*) became the gold standard of the Enlightenment's 'systematic spirit' (*esprit systématique*). Newton's method does not proceed from concepts and axioms to phenomena, but the other way around: it starts from phenomena to deduce laws of nature, to 'discover ... regularity in the phenomena themselves, as the form of their immanent connection'.[49]

This is evidently a neo-Kantian reading of the philosophy of the Enlightenment, a way of linking the philosophical roots of the Kantian *regulative* demand to the *philosophes'* conception of 'reason' and to their admiration for Newton's method. From this perspective, Newton's method of deduction from phenomena comes to fulfil a regulative task: it unfolds lawlikeness as immanent in phenomena.

These ideas were further spelled out in Cassirer's later book *Determinism and Indeterminism in Modern Physics*, where Cassirer reaffirmed and strengthened the theses already expressed in *Substance and Function* in the light of the recent scientific developments. According to Cassirer, modern physics has not given up the salient features of Kant's philosophical project. On the contrary, quantum mechanics has only made evident the fact that Kant's philosophical apparatus is to be thought of not as rigid but as dynamic. Hence:

---

[48] Cassirer (1932). English translation (1962), p. 5.  [49] *Ibid.*, p. 9.

the *a priori* that can still be sought and that alone can be adhered to must do justice to this flexibility. It must be understood in a purely methodological sense. *It is not based on the content of any particular system of axioms, but refers to the process whereby in progressive theoretical research one system develops from another.*[50]

The highest expression of Cassirer's advocacy of systematicity can be found in the architectonic of scientific knowledge that he portrayed as consisting of (1) results of measurements, (2) laws, and (3) principles. Cassirer made it clear that this distinction should not be read hierarchically, or as implying some sort of reductionism. It is rather a purely 'architectonic' distinction, so to speak. The necessity of each law is a function of its place within this architectonic, whereby results of measurements are 'the alpha and omega of physics, its beginning and end'. On the other hand, the role of scientific principles consists in providing a 'synopsis', in conferring an inner organization on scientific knowledge. As such, scientific principles do not prescribe the way nature is to be, but rather the form of the laws according to which we should order phenomena. Thus, scientific principles are not on a par with empirical laws: they are rather 'the birthplace of natural laws, a matrix as it were, out of which new natural laws may be born again and again ... The principles of physics are basically nothing but such means of orientation, means for surveying and gaining perspective ... they teach us how to find the direction in which we have to advance.'[51]

Kant's regulative principle of systematicity, as a necessary condition for identifying empirical regularities as lawlike, finds its natural expression in Cassirer's architectonic of scientific knowledge. Results of measurement and scientific principles occupy the two complementary poles of this architectonic. The former provide the empirical basis. The latter fulfil the regulative task of systematizing and conferring an order on the empirical manifold, as an integral and indispensable part of empirical knowledge itself.

---

[50] Cassirer (1936), English translation (1956), p. 74, emphasis added. Cassirer's interpretation of Einstein's equivalence principle is symptomatic of this (regulative in spirit) reinterpretation of the 'a priori'. According to Cassirer, the principle of equivalence would not play the role of a constitutive a priori principle, but of a systematizing principle, i.e. of a principle that insofar as it is able to give new form and organization to well-known empirical facts, it accomplishes a regulative task: 'The equality of the gravitational and inertial mass of a body, which had long been known and established as a fact of experience, was placed in a completely new light conceptually by the general theory of relativity. As Einstein expressed it, this equality was not only registered but interpreted. In this continual oscillation the essential character of the dynamic form is established and verified as a form that not merely receives and incorporates new material passively but seeks the new material, and because it seeks it, it is able to form and organise it.' *Ibid.*, p. 75.

[51] *Ibid.*, pp. 52–4.

In the light of this architectonic of knowledge, the concept of physical reality also gets reinterpreted. Traditionally, in what Cassirer calls the 'substantialistic conception', it is assumed that there are entities bearing certain properties and entering into definite relations with other entities that can be expressed by laws of nature. Laws of nature are read off the entities, their properties and relations. From Cassirer's 'functional' viewpoint, on the other hand, entities no longer constitute the self-evident starting point. They are not the *terminus a quo* but the *terminus ad quem* of scientific inquiry:

> The extent of the dominance of these laws marks the extent of our objective knowledge. Objectivity or objective reality, is attained only because and insofar as there is conformity to law – not vice versa ... Apart from this reality there exists for us no other objective reality to be investigated or sought after.[52]

This is perhaps the most important lesson of Cassirer's view. And it has far-reaching consequences for current debates in philosophy of science. It is onto this philosophical tradition that I intend to latch my analysis of the exclusion principle.

## 1.5 The exclusion principle: a Kantian perspective

I have lingered on the Kantian tradition in the preceding sections to set the stage for the philosophical analysis of the exclusion principle that I shall carry out in this book. This analysis owes an obvious debt to Buchdahl's interpretation of Kant as well as to Cassirer in regarding the necessity and lawlikeness of the Pauli principle as a result of the regulative function it played in quantum mechanics. The historical process that goes from the humble origins of Pauli's principle as a phenomenological rule in spectroscopy to its subsequent embedding into a growing theoretical framework is symptomatic of the necessity and special nomological status that accrued to it in virtue of the specific place and role it came to play in that framework.

The upshot of this monograph is to show that an empirical and contingent rule such as Pauli's 1924 exclusion rule attained lawlikeness and necessity because of the *systematizing* role it accomplished in the quantum mechanics framework. As the discussion in the previous sections aimed to

---

[52] *Ibid.*, p. 135.

highlight, systematization as a Kantian regulative demand should here be understood not as a desirable, albeit additional and hence dispensable, feature of scientific inquiry, but rather as underpinning the possibility itself of identifying empirical regularities as lawlike and regarding them as necessary laws.

The exclusion principle has humble origins. It was introduced as a phenomenological rule in the early 1920s, when some spectroscopic anomalies known since the end of the nineteenth century proved persistent to the extent of putting the old quantum theory in crisis. Some insight in atomic physics was required. At the end of 1924, the introduction of Pauli's exclusion rule and of the concept of electron spin that accompanied it finally provided physicists with the long-sought missing element of this jigsaw. What happened in late 1924 was a genuine revolution. The standard spectroscopic model of the time (the so-called *atomic core model*) that had been in vogue among Landé, Bohr, Heisenberg, and Pauli himself for many years, was finally dismissed. It was Pauli who first abandoned the atomic core model and introduced a fourth degree of freedom for the electron. Without giving any mechanical interpretation of it in terms of a 'spinning' electron (as Uhlenbeck and Goudsmit were to do a year later) Pauli spoke of a mechanically non-describable *Zweideutigkeit* (twofoldness) of the electron angular momentum.

This was the turning-point that marked the scientific revolution around 1925, where old concepts fell short of grasping new phenomena, and new concepts had to be tentatively introduced. Pauli's rule emerged during the revolutionary transition from the old quantum theory to the new quantum theory, which involved the dismissal of the atomic core model, the introduction of a new crucial physical property (the electron's spin), and its subsequent modelling according to an increasing degree of abstraction (from the semi-classical spinning electron model to Dirac's relativistic equation for the electron in 1928). It was introduced as a tentative phenomenological rule, while opinions diverged about the nature of the fourth degree of freedom that the new rule presupposed, as I shall reconstruct in Chapter 2.

The continuity that Friedman envisages as necessary for the prospective rationality of scientific revolutions does not reside in this case in the principle itself, or in a well-known empirical fact elevated to the status of a scientific principle. Continuity should here be looked for in the experimental and theoretical *milieu* from which Pauli's rule was derived, and in the gradual process of transformation of the old quantum theory due to its increasing inability to deal with anomalous phenomena. It is this piecemeal transformation of the old quantum theory that ultimately accounts for the prospective rationality and even intelligibility of this revolutionary transition, as I shall argue in Chapter 3 against Kuhn's incommensurability thesis. Pauli's rule was one of the most important products of this revolutionary transition: it was 'deduced from phenomena' with the help of some theoretical assumptions of the old quantum theory (Section 3.3.1). Following Cassirer, I shall appeal to Newton's method of deduction from phenomena as a scientific method unfolding the lawlikeness immanent in phenomena. Pauli's rule was deduced from phenomena precisely in this spirit. The theoretical premises with the help of which this deduction from phenomena proceeded were on rather shaky grounds: they were merely phenomenological laws, methodological guidelines, or classical theorems of electrodynamics that appeared to be violated.

The shaky grounds of these theoretical premises and the revolutionary context could not provide Pauli's rule with the necessity and the special nomological status it has nowadays. It was only a phenomenological rule on a par with many others composing that hodgepodge of numerology and rules of thumb so typical of the old quantum theory. Those who really accomplished the revolution and took the first step to transform Pauli's rule into a scientific principle were Fermi and Dirac. It was only when the rule was transplanted from the original spectroscopic context onto the domain of the rising quantum mechanics that *Pauli Verbot* acquired a new significance. It became the manifestation of the antisymmetric character of the state function of an assembly of indistinguishable particles. The 'prohibition rule' became the exclusion principle. In his pioneering 1926 contribution, Dirac referred to it as Pauli's exclusion principle; and it was more than a terminological baptism. It was indeed the reformulation in terms of Fermi–Dirac statistics that allowed Pauli's rule to transcend its humble origins in atomic spectroscopy and extend its range of nomological applicability. As I shall describe in Chapter 4, necessity and nomological strength accrued to Pauli's rule in virtue of its progressive embedding into a growing theoretical framework (from the non-relativistic quantum mechanics of the magnetic electron, to relativistic quantum field theory within which the spin–statistics theorem was proved in 1940), and in virtue of the *systematization* it accomplished in the new framework, as is to be expected on a dynamics of reason along the Buchdahl/Cassirer lines.

As we shall see in Chapter 5, the principle, with its enlarged nomological status given by the spin–statistics theorem, could be later applied to particle physics in a heuristically fruitful way. Indeed, Fermi–Dirac statistics turned out to be one among many types of quantum statistics (not only Bose–Einstein, but also para-Fermi and para-Bose) allowed by permutation invariance. Thanks to this further embedding into the larger mathematical structure disclosed by permutation invariance, and thanks to the discovery of some recalcitrant evidence for the exclusion principle in the quark theory, two rival research programmes were articulated in the 1960s. One accommodated the negative evidence by revoking the strict validity of the exclusion principle and assuming the existence of Pauli-violating spin-1/2 particles (paraparticles). The other programme shielded the exclusion principle from recalcitrant evidence by introducing the auxiliary assumption of 'colour' as a further degree of freedom for quarks, and in this way led to a non-Abelian gauge theory of strong interaction: quantum chromodynamics.

The comparison between these two programmes in Chapter 5 will provide us with an excellent foil to reassess the Duhem–Quine thesis, and Quine's epistemological holism. I shall argue that the enduring nomological validity of the exclusion principle is not due to either its playing a constitutive function or its being a well-entrenched principle, but rather to its being empirically well-supported. More precisely, the negative experimental results found in the 1990s in the context of paronic theories[53] for Pauli-violating particles, and, on the other hand, the wide-ranging array of experimental evidence for the auxiliary assumption of 'colour' for quarks, have both offered evidential warrant for the strict validity of the principle. In the end, the analysis of the lawlikeness of the exclusion principle that I shall propose does justice to the revisable and experimentally testable nature of the principle in as much as it grounds its enduring nomological validity on the degree of empirical support that it has received.

---

[53] As we shall see in Section 5.3.1.1, in 1989 Greenberg and Mohapatra introduced 'parons' as hypothetical particles violating the exclusion principle by a small amount. This allowed for the first time the parametrization and measurement of eventual violations of the principle, and hence opened the door to experimental tests in the 1990s.

# 2

# The origins of the exclusion principle: an extremely natural prescriptive rule

The exclusion principle was the final outcome of Pauli's struggle to understand some spectroscopic anomalies in the early 1920s: doublets were observed in the spectra of alkali metals, singlets and triplets in the spectra of the alkaline earths, and even more anomalous patterns were observed when chemical elements were placed in an external magnetic field (anomalous Zeeman effect and Paschen–Back effect). These anomalous spectra challenged the old quantum theory, and prompted a radical theoretical change (Section 2.1). From 1920 to 1924 Alfred Landé, Werner Heisenberg, and Niels Bohr were all engaged in trying to save the traditional spectroscopic model (the so-called atomic core model) and to reconcile it with the observed anomalies. The impasse was solved only with Pauli's introduction of a fourth degree of freedom for the electron, and the consequent demise of the atomic core model (Section 2.2). What Pauli called the 'twofoldness' [*Zweideutigkeit*] of the electron's angular momentum was soon reinterpreted as the electron's spin (Section 2.3). Pauli's exclusion rule was announced in this semi-classical spectroscopic context that characterized the revolutionary transition from the old quantum theory to the new quantum theory around 1925.

## 2.1 The prehistory of Pauli's exclusion principle

### 2.1.1 Atomic spectra and the Bohr–Sommerfeld theory of atomic structure

The existence of spectral lines had been known to scientists since the beginning of the nineteenth century when Wollaston and Fraunhofer first observed the dark absorption lines in the spectrum of the Sun. It took almost eighty years – the time necessary to improve the quality of technical instruments – to find an empirical law governing the line distribution in the hydrogen spectrum as observed in stars. In 1885 Balmer discovered that each of the fourteen lines composing the spectrum of hydrogen obeyed the simple rule

$$\lambda = h \frac{n_2^2}{n_2^2 - n_1^2} \tag{2.1}$$

where $\lambda$ is wavelength, $h = 3645.6$ Å, $n_1 = 2$ and $n_2 = 3, 4, 5, 6, \ldots$ respectively for the first, second, third, and so forth elements of the series. A few years later it became clear that this formula could be derived from another more general formula due to Rydberg, which gave the frequency for any spectral series as

$$\nu_n = \frac{R}{(n_1 + \mu_1)^2} - \frac{R}{(n_2 + \mu_2)^2} \tag{2.2}$$

where $\nu_n$ is the frequency of the $n$-th member of the series, $R$ is the Rydberg constant, and $\mu$ denotes spectral terms (e.g. $\mu_1 = {}^3P_1, \mu_2 = {}^3S_1$). By setting $\mu_1 = 0$, $\mu_2 = 0$, $n_1 = 2$ and $n_2 = 3, 4, 5, \ldots$ Rydberg's formula reduces to Balmer's.

Rydberg was the first to distinguish between a *sharp series* (S) and a *diffuse series* (D). Other types of series were later discovered: the so-called *principal series* (P) and the *fundamental series* (F). Jointly they form the *four chief series* (S, P, D, F) available for every type of line (i.e. singlet, doublet, triplet, ...).[1] According to Rydberg's formula, each line of a series is then given by the difference between two spectral terms.[2] This was a useful phenomenological law, but did not offer a theoretical understanding of spectra. Only with Bohr's atomic theory of hydrogen in 1913, was the first step towards such an understanding taken.

Not only could Bohr derive from his atomic theory Rydberg's and Balmer's formulae for the hydrogen series, but he provided Rydberg's

---

[1] The standard notation uses the superscript on the left of the capital letter to denote the type of line; for instance the singlet four chief series are denoted as ${}^1S$, ${}^1P$, ${}^1D$, ${}^1F$.

[2] For instance, the three lines composing a triplet of the sharp series S are given by the following differences: ${}^3P_2 - {}^3S_1$; ${}^3P_1 - {}^3S_1$; ${}^3P_0 - {}^3S_1$, where the subscripts indicate the specific line of the series at issue (e.g. ${}^3P_2$ denotes the second line of the principal series triplet). These differences are obtained by plugging into Rydberg's formula $\mu_2 = {}^3S_1$ (fixed term) and $\mu_1 = {}^3P_2$, ${}^3P_1$, ${}^3P_0$ respectively (running terms).

phenomenological law with a physical meaning: the difference between the spectral terms was interpreted as the difference between two energy-states (stationary states) of the atom. The frequency of a spectral line $\nu_n$ would then be proportional to the difference between the initial state $W_2$ in which the electron is located and the final state $W_1$ into which it jumps by emitting radiation. The fixed term $\mu_2$ of Rydberg's formula was identified with the final or lower energy-state, and the running term $\mu_1$ with the different possible initial states. From Planck's quantum postulate it follows that radiation is not emitted continuously but only in discrete quanta of energy, so Bohr wrote his formula $h\nu = W_2 - W_1$, where $h$ is Planck's constant.

Sommerfeld significantly extended Bohr's atomic theory of hydrogen and hydrogen-like atoms to include elliptic orbits.[3] In his model, an electron moving in an elliptic orbit is a system with two degrees of freedom and two quantum conditions. In polar coordinates, the position of the electron is given by the electron–nuclear distance $r$ and the azimuthal angle $\varphi$ between $r$ and the major axis of the elliptic orbit. Accordingly the electron had two momentum coordinates:

$$\text{(i)} \quad p_\varphi = mr^2 \dot{\varphi} \tag{2.3}$$

$$\text{(ii)} \quad p_r = m\dot{r} \tag{2.4}$$

where $m$ is electron mass. By applying Bohr's quantum condition to the two momentum coordinates, Sommerfeld's quantum conditions followed:[4]

$$\text{(1)} \quad \oint p_\varphi \, d\varphi = kh \tag{2.5}$$

$$\text{(2)} \quad \oint p_r \, dr = rh \tag{2.6}$$

where $k$ and $r$ are respectively the *azimuthal* and the *radial quantum number* and, following Bohr's quantum condition, they take only discrete integral values ($r = 0, 1, 2, 3, \ldots$; $k = 1, 2, 3, 4, \ldots$). The sum $r + k$ was equal to $n$, where $n$ was the so-called *principal quantum number* denoting the energy-state of the electron. Like $r$ and $k$, $n$ also took only integral values ($n = 1, 2, 3, \ldots$). While $r$ and $k$ determined respectively the size and shape of the orbit, in order to determine the orientation of the orbit in an external magnetic field, it was necessary to introduce a third degree of freedom.

---

[3] Sommerfeld (1916a), (1919).   [4] For details see White (1934), pp. 42ff.

As is well known from classical electrodynamics, in particular from Larmor's theorem,[5] in the presence of an external magnetic field of intensity $H$ the orbit of the electron precesses about the field direction with uniform angular velocity

$$\omega = H\frac{e}{2mc} \qquad (2.7)$$

and uniform frequency

$$\nu_\text{L} = H\frac{e}{4\pi mc} \qquad (2.8)$$

This is equivalent to saying that Larmor's angular velocity is given by the field strength $H$ times the magneto-mechanical ratio $\mu/p = e/(2mc)$, where $\mu$ is the electron's magnetic moment due to its orbital motion and $p$ is its orbital angular momentum; on the other hand, Larmor's precession frequency is given by $H/2\pi$ times $\mu/p$. Interestingly enough, Larmor's magnetic effect was known to be proportional to the electron's charge-to-mass ratio $e/m$ well before Thomson's discovery of the electron. The unit $eh/(4\pi mc)$ (or, equivalently, $e\hbar/(2mc)$) was called the Bohr magneton $\mu_\text{B}$ and in the old quantum theory it was taken as the unit of the magnetic moment for an electron bound in an atom. According to Larmor's theorem, the presence of an external magnetic field did not alter the size and shape of the electron's orbit, because of a balance between the Coriolis force and the Lorentz radial force on the electron. But it altered the orientation of the orbit with respect to the field direction. If we represent the orbital angular momentum of the electron by a vector **K** (in modern notation **L**), under the effect of Larmor precession **K** turns about the field vector **H** with an angle $\theta$. According to Larmor's theorem, the angle $\theta$ is constant, i.e. **K** takes continuous values in the precession. But, as Sommerfeld[6] pointed out, the vector **K** did not project itself on the field direction in a continuous way. Rather, the cosine of $\theta$ took only discrete values denoted by a new quantum number, the so-called *magnetic quantum number m*, which like $n$ and $k$ took only discrete integral values[7] ($m = \pm 1$, $\pm 2$, $\pm 3$, ..., $\pm k$ for $k = 1, 2, 3, 4, ..., n$). For instance, for $n = 1$ and $k = 1$, $m$ took the values $\pm 1$; for $n = 2$ and $k = 1, 2$, $m = \pm 1, \pm 2$ corresponding to the four possible projections of the orbital angular momentum vector **K** on the field direction (Fig. 2.1): the orbit appeared to be

---

[5] Larmor (1897).    [6] Sommerfeld (1916b).
[7] It was assumed that $m$ could not take the value 0, because this would lead to the electron colliding with the nucleus.

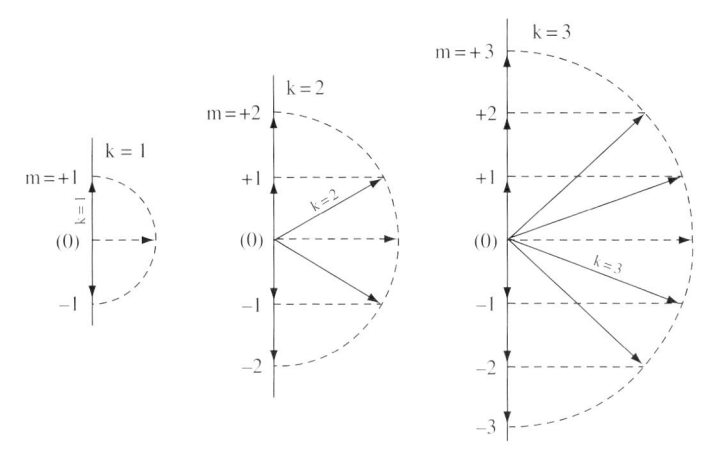

## 2.1 The prehistory of Pauli's exclusion principle

Fig. 2.1 Space-quantization diagrams for Bohr–Sommerfeld orbits with orbital angular momentum $k = 1$, $2$, and $3$.

*Source*: H. E. White (1934) *Introduction to Atomic Spectra* (New York, London: McGraw-Hill Book Company). Reproduced with permission of The McGraw-Hill Companies.

space-quantized. And space quantization seemed to be confirmed by the Stern–Gerlach experiment in 1921.[8]

Following up on Sommerfeld, at the beginning of the 1920s Bohr developed a building-up schema for the elements of Mendeleev's periodic table. He assumed that the number of electrons in an atom was equal to the atomic number $Z$; each electron was in a definite quantized state and no radiation was emitted as long as the electron remained in its stationary state. Each stationary state corresponded to a possible $n_k$-orbit for the electron, where $n$ was the principal quantum number and $k$ the azimuthal quantum number. Bohr fixed $n = 1$ for the electrons of hydrogen ($Z = 1$) and helium ($Z = 2$); $n = 2$ for the electrons of the elements from lithium

---

[8] The Stern–Gerlach experiment consisted of directing a beam of silver atoms in a high vacuum through collimating slits along a strong field gradient so that the beam split into two sub-beams. According to classical mechanics, since the magnetic moments of the atoms were isotropically distributed (i.e. they had equal probability to take on any value between $M$ and $-M$), one would expect one single spot at the centre of the distribution corresponding to the average magnetic moment. But Stern and Gerlach observed two spots centred at two different points corresponding to two discrete values of $M$. This result was greeted as a confirmation of Sommerfeld's space quantization, as evidence for the discrete orientations atoms would take up in the presence of an external magnetic field. As a matter of fact, the Stern–Gerlach experiment can be unambiguously interpreted only with the introduction of the spin: the correct explanation is that the total angular momentum of the silver atoms coincides with the spin of the valence electron (which is alone responsible for the magnetic moment $M$ of the atom). And the spin projection takes only two discrete eigenvalues ($+1/2$ and $-1/2$ in units of $h/2\pi$) corresponding to the two observed spots.

($Z=3$) to neon ($Z=10$); $n=3$ for the electrons of the atoms of the 3rd period, and so forth. Thus, hydrogen would have a single $1_1$ orbit ($n=1$ and $k=1$); helium would have two $1_1$ orbits for its two electrons; lithium would have the two $1_1$ orbits of helium plus a new $2_1$ orbit for its 3rd electron; the 4th electron of beryllium ($Z=4$) would be in a second $2_1$ orbit, and so on up to carbon ($Z=6$), whereas the 7th electron of nitrogen ($Z=7$) to the 10th electron of neon ($Z=10$) would be placed in many $2_2$ orbits.

Each chemical element then seemed to be built up by adding an outer electron to the electronic configuration of the previous element in the periodic table, which consequently became the atomic core of the element at issue. Bohr's fundamental assumption was that the addition of the outer electron did not affect the quantum numbers $n$ and $k$ of the already bound electrons of the atomic core. This invariance and permanence of quantum numbers[9] is known as Bohr's building-up principle (*Aufbauprinzip*). Since this principle will be crucial for the rest of our story, it is necessary to linger a little on it.

The building-up principle was justified by the classical theory of conditionally periodic systems via Bohr's correspondence principle. The correspondence principle established a general correspondence between classical electrodynamics and Bohr's atomic theory. In the original formulation, the principle established a correspondence – in the region of high quantum numbers – between the classical concept of harmonic components in the motion of an electrically charged particle and the quantum concept of an electron's transitions between stationary states.[10] However, Bohr soon regarded this principle, more generally, as the methodological guideline underpinning his project to shape quantum theory as closely as possible along the lines of classical physics.[11] As such, the correspondence principle is at work in the justification of the building-up principle: via the

---

[9] See Bohr (1923a). Reprinted in Bohr (1977), p. 632.

[10] 'Although the process of radiation cannot be described on the basis of the ordinary theory of electrodynamics, according to which the nature of the radiation emitted by an atom is directly related to the harmonic components occurring in the motion of the system, there is found nevertheless to exist a far-reaching correspondence between the various types of possible transitions between the stationary states on the one hand and the various harmonic components of the motion on the other. This correspondence is of such a nature that the present theory of spectra is in a certain sense to be regarded as a rational generalization of the ordinary theory of radiation.' Bohr (1914). Reprinted in Bohr (1976), p. 301.

[11] For a detailed analysis of the implications of the correspondence principle for Bohr's later research as well as for the development of quantum mechanics, see Petruccioli (1988).

## 2.1 The prehistory of Pauli's exclusion principle

correspondence principle, Bohr extended analogically the classical notion of adiabatic invariants to atomic structure.

From the end of nineteenth century it was known that all periodic mechanical systems possessed adiabatic invariants, i.e. magnitudes that remained invariant while certain parameters underwent slow variations.[12] Bohr extended in a non-mechanical way the classical notion of adiabatic invariants to the electrons of an atom. In the case of a single-electron atom such as hydrogen, the action of the electron moving around the nucleus was taken to be adiabatically invariant: the atom was unaffected by the slowly varying forces of the electric field. But problems arose with many-electron atoms, because in this case the motion of the electrons did not exhibit periodic properties. It was at this point that Bohr resorted to the correspondence principle. Even if it was not possible to apply the classical theory of conditionally periodic systems to $n_k$-orbits of many-electron atoms, yet

> the general stability of atomic structures leads to the view that, in every atom with several electrons, there are properties of the motion of each electron and of the interplay among the electrons, which possess an invariant character that cannot be explained mechanically, but that finds its meaningful expression just in the quantum numbers. This view might be called a formal postulate of the invariance and permanence of quantum numbers.[13]

As a result, the statistical weights of the stationary states characterized by these quantum numbers were now regarded as adiabatically invariant. In other words, the addition of an electron in the building-up process was expected to let the atom undergo a slow variation of the electric field, without however affecting the stationary states of the already bound electrons. This explains why the building-up principle can also be known in the literature as Bohr's *adiabatic principle*.[14] But this terminology is misleading given the existence of another adiabatic principle in quantum theory due to Ehrenfest;[15] henceforth I prefer to refer to it as the building-up principle.

---

[12] The concept of adiabatic change goes back to Boltzmann's and Clausius's attempts to reduce the second law of thermodynamics to pure mechanics, while the notion of adiabatic motion was developed by Hertz and Helmoltz.

[13] Bohr (1923a). Reprinted in Bohr (1977), pp. 631–2.

[14] See for instance Serwer (1977), p. 204.

[15] Ehrenfest (1913) was the first to extend the classical notion of adiabatic invariants to quantum theory as well as the first to introduce – following a terminology suggested by Einstein – an adiabatic principle. For the history of this principle, see Jammer (1966), second edition (1989), pp. 99–107. It is worth noticing that Bohr derived the invariance of statistical weights precisely from Ehrenfest's adiabatic principle: 'A basis for the

Bohr's building-up schema offered a relatively simple and elegant classification of electrons in $n_k$-orbits, in which symmetry considerations played a relevant role.[16] Yet the schema had a major drawback: symmetry considerations about the electronic distribution often spoiled the consistency of the building-up process through the periodic table. For example, argon ($Z = 18$) was built up from the atomic core of the previous noble gas in the periodic table, namely neon ($Z = 10$), by adding eight new electrons – placed in suitable $3_k$-orbits – to the $2_k$-orbits of neon. Then, one would expect the next noble gas, krypton ($Z = 36$), to be built up from the atomic core of argon by adding eighteen new electrons – placed in suitable $4_k$-orbits – to the $3_k$-orbits of argon. But this was not the case. According to Bohr's building-up schema, the electronic distribution of argon was not strictly maintained in the building-up process of krypton. Argon orbits were instead reopened and the electronic distribution rearranged so as to satisfy symmetry in the structure of krypton, i.e. so as to have three groups of six electrons each, placed in suitable $3_k$-orbits:

$$\mathrm{Kr}(Z = 36) = (\mathrm{neon})^{10}\ (3_1)^6\ (3_2)^6\ (3_3)^6\ (4_1)^4\ (4_2)^4.$$

Analogous arbitrary reopenings of orbits occurred in the 4th and 5th period. As we shall see, this drawback led Edmund Stoner, a few years later, to modify Bohr's building-up schema in a way that would turn out to be crucial for Pauli's introduction of his exclusion rule. Suffice it to say that despite this drawback, Bohr's building-up schema led to the successful prediction of the chemical properties of an undiscovered element with atomic number 72, and Hevesy and Coster's discovery of hafnium in 1922 was hailed as a confirmation of Bohr's schema.[17] The widespread consensus that Bohr's theory gained in the period up to 1920 could not however prevent it from facing serious difficulties when it came to the explanation of some puzzling spectroscopic anomalies.

---

treatment of this problem [i.e. the problem of the weights that must be assigned to the individual stationary states] is contained in a general condition, derived by Ehrenfest ... stating that in a continuous transformation of a system, the statistical weights of the individual stationary states do not change as long as the degree of periodicity remains the same. For systems for which the degree of periodicity is equal to the number $s$ of the degrees of freedom, this condition involves that all stationary states must be assigned the same weight $h^s$.' Bohr (1923a). Reprinted in Bohr (1977), p. 616.

[16] This is particularly evident in the electronic distribution of neon ($Z = 10$), where the ten electrons were distributed as follows: the first two electrons in two ($1_1$) orbits, the following four electrons in four ($2_1$) orbits and the last four electrons in four ($2_2$) orbits. For a detailed discussion of Bohr's schema and its drawback, see Heilbron (1982), p. 273.

[17] For the discovery of hafnium, see Kragh (1979).

### 2.1.2 The doublet riddle and the riddle of statistical weights

Bohr had resorted to the principal and azimuthal quantum numbers to distribute electrons in $n_k$-orbits. However, it soon became evident that these two quantum numbers were not sufficient to account for the multiplets observed in the spectra of many-electron atoms. In 1920 doublets and triplets were well known to spectroscopists, and after the work of Catalan in 1922 higher multiplicities were discovered. The jungle of observed doublets, triplets, quintuplets, and sextuplets in the series spectra hinted at the existence of a multiplicity in the energy-levels that was not adequately mirrored by the values of $n$ and $k$ alone. Multiplets stemmed supposedly from an energy difference due to a splitting of the orbital angular momentum state $k$. Thus, energy-levels for individual electrons labelled s, p, d, f[18] were associated with the azimuthal quantum numbers $k = 1, 2, 3, \ldots n$, and it was assumed that each energy-level – except s – could further split to yield doublets, triplets, or other multiplets. For instance, the series spectra of alkali metals, i.e. the elements of the first column of the periodic table (Li, Na, K, Rb, Cs), exhibited doublets that supposedly corresponded to double energy-levels p, d, f. The spectra of the alkaline earths, i.e. the elements of the second column (Be, Mg, Ca, Sr, Ba), could display triplets and singlets, but only singlets were usually observed. How did the alkali doublets originate? And why did alkaline earths exhibit only singlets, even if triplets were theoretically allowed? More generally, why did the angular momentum state split?

The current physical explanation of multiplets involves the so-called spin–orbit coupling. To put it in an extremely simple way, consider the semi-classical vector model of a one-electron atom, where we can represent the total angular momentum as a vector **J**, which is the sum of the valence electron's orbital angular momentum vector **L** and of its spin angular momentum vector **S**. The electron moves in the electrostatic field **E** created by the protons of the nucleus. This field appears in the electron frame as an internal magnetic field **B** at the electron normal to the plane of the orbit. The magnetic moment of the electron spin interacts with this internal field in such a way that it can line up either parallel to **B** when $\mathbf{J} = \mathbf{L} - \mathbf{S}$, or antiparallel to **B** when $\mathbf{J} = \mathbf{L} + \mathbf{S}$. These two possible spin–orbit

---

[18] For instance $k = 1$ corresponds to s; $k = 2$ to p; $k = 3$ to d and $k = 4$ to f. Notice the notational difference between the small letters s, p, d, f referring to the energy-levels (and hence to the spectral terms) for individual electrons and the capital letters S, P, D, F referring to the spectral terms resultant from the combination of any two or more non-equivalent s, p, d, f electrons.

combinations correspond to doublet fine structure levels (e.g. for p electrons in sodium and potassium, $^2P_{1/2}$ and $^2P_{3/2}$). It is the spin–orbit coupling that is at work in the doublet shift of the alkalis, as well as in other spectroscopic anomalies. No wonder then that in the pre-spin era before 1925 the alkali doublets appeared as a puzzling phenomenon.

The first step towards the solution of this riddle came from the analysis of the fine structure of X-ray spectra. As known since 1914,[19] an atom emits X-ray doublets during the transition of an outer electron into an internal vacant orbit. Sommerfeld[20] pointed out that the width of the X-ray doublets was proportional to the fourth power of the effective nuclear charge, $(Z-s)^4$, where $Z$ is the atomic number corrected by a factor $s$ due to the screening exerted by the inner electrons of the atom. Sommerfeld thought that the electron responsible for the emission of X-rays underwent a relativistic increase in mass due to the different orbital angular momenta of the two orbits having different eccentricities. This difference accounted satisfactorily for the X-ray doublet fine structure. The step from the 'relativistic' X-ray doublets to the 'optical' alkali doublets was short, and at the beginning of the 1920s attempts were made to extend the relativistic explanation to optical doublets.[21] But the two kinds of doublets were not analogous, since the energy-levels at work in the alkali doublets did not have different orbital angular momenta. No unified explanation was possible for relativistic and optical doublets: hence the so-called 'doublet riddle'.[22]

The second important step was taken once again by Sommerfeld a few years later.[23] To bring about the required multiplicities in the energy-levels, Sommerfeld introduced a further quantum number $j$ in addition to Bohr's $n$ and $k$. This new quantum number, called the *inner quantum number*, denoted the energy-sublevels into which the $n_k$-orbits supposedly split to give rise to multiplets. Without any clear geometrical or physical meaning, $j$ led nonetheless to a useful empirical rule for doublets and triplets: doublets were supposed to originate from two energy-sublevels $j=k$, $k-1$ while triplets from three sublevels $j=k$, $k-1$, $k-2$, with $j$ selection rule $\Delta j = 0, \pm 1$. One year later Landé[24] provided the inner quantum number $j$ with a physical meaning in the light of the so-called *atomic core model*.

---

[19] For a historical reconstruction of the research on the X-rays and their spectra see Heilbron (1966), (1967).
[20] Sommerfeld (1916a). [21] See de Broglie and Dauvillier (1922).
[22] See Forman (1968). [23] Sommerfeld (1920). [24] Landé (1921a), (1921b).

## 2.1 The prehistory of Pauli's exclusion principle

The atomic core model – differently re-elaborated by Landé, Heisenberg, Bohr and Pauli – became the standard spectroscopic model for the explanation of multiplets until Pauli's breakthrough in 1924. According to the model, the atomic core of any atom (i.e. the internal part – valence electrons aside – consisting of the nucleus plus the bound electrons in closed shells) was supposed to have a non-zero angular momentum in units of $h/2\pi$, denoted by a core quantum number $r$ and represented by the core angular momentum vector **R**. The orbital angular momentum of the valence electron, denoted by the azimuthal quantum number $k$, was represented by the vector **K** (in modern notation **L**). The sum of **K** and **R** was supposed to give the total angular momentum vector of the atom **J**. Landé identified the total angular momentum with Sommerfeld's inner quantum number $j$. He was perhaps driven to such an identification by Sommerfeld's selection rule for $j$, which surprisingly coincided with Rubinowicz's selection rule for the total angular momentum of the atom.[25] Multiplets seemed to arise from different energy-sublevels due to different orientations of **K** with respect to **R**, and, as required by space quantization, only a few discrete orientations (in units of $h/2\pi$) were allowed, namely those associated with the multiplet components.

The atomic core model accounted satisfactorily for the observed multiplets, because the supposed electron–core coupling was a mistaken empirical surrogate for the spin–orbit coupling: it entailed the same phenomena that are actually entailed in the spin–orbit coupling. The atomic core soon turned out to play no role in the total angular momentum of the atom and its alleged angular momentum was substituted by the electron's spin angular momentum. This turning point dates back to Pauli's spectroscopic research, as I shall explain in Section 2.2.4. But in the pre-spin era, in order to account for the observed multiplicity of spectral terms, the most natural way was to ascribe a non-zero angular momentum to the atomic core and to assume a coupling between the core and the electron's orbital angular momentum.

---

[25] Rubinowicz (1918). Rubinowicz was one of Sommerfeld's students and assistants. While working on the conservation of angular momentum in the classical process of radiation, he found the following integral selection rules for the azimuthal and the magnetic quantum numbers: $\Delta k = 0, \pm 1$ and $\Delta m = 0, \pm 1$. But Rubinowicz identified the azimuthal quantum number with the total angular momentum of the atom, rather than with the electron's orbital angular momentum. Accordingly, the physical meaning of this rule was that the total angular momentum of the atom was quantized as required by Sommerfeld's space quantization. For the historical reconstruction of the way Landé was driven to identify $j$ with the total angular momentum via Rubinowicz's selection rule, see Forman (1970), p. 196.

There were however a few striking difficulties with the atomic core model. First, it contradicted Bohr's building-up principle. Consider, for instance, the atomic cores of alkalis. They coincide with the noble gases occupying the column immediately preceding them in Mendeleev's periodic table. Hence, if alkalis had a non-zero core momentum as the model required, noble gases too should have a non-zero angular momentum. But this was in contrast with experimental evidence.

Furthermore, even granted a non-zero angular momentum for noble gases, it was still impossible to get doublets and, more generally, even multiplets. According to the atomic core model, the values of the inner quantum number $j$ were defined as $|k-r| \leq j \leq |k+r|$. But given the integral values of the electron azimuthal quantum number $k$ and the core quantum number $r$, it was possible to get spectral terms of odd multiplicity (e.g. singlets, triplets, ...), but not of even multiplicity (e.g. doublets, quadruplets, ...). Alkali doublets remained unexplained.

The atomic core model contradicted Bohr's building-up principle in another way: the inner quantum number $j$ and the number of states associated with it did not remain adiabatically invariant during the building-up process. As Landé[26] pointed out, the inner quantum number $j^+$ of an ion did not coincide with the core quantum number $r$ of the immediately preceding element in the periodic table, as Bohr's principle required. Rather, Landé found that

$$j^+ = r - 1/2 \qquad (2.9)$$

and more generally

$$j = r + k - 1/2, r + k - 3/2, \ldots \qquad (2.10)$$

The number of states associated with $j$ seemed to decrease in the building-up process by an inexplicable half-integral unit. And this decrease brought with it a corresponding decrease in the statistical weights of the states. But the building-up principle required the statistical weights to remain invariant. The riddle of statistical weights[27] was no less puzzling than the doublet riddle. Indeed they were two sides of the same coin as the analysis of the Zeeman effect soon revealed.

---

[26] Landé (1923a) See also Serwer (1977), pp. 206–7.
[27] For this terminology see Serwer (1977), pp. 204–7.

### 2.1.3 The anomalous Zeeman effect and the mystery of half-integral quantum numbers

The atomic core model was also deployed to account for an even more puzzling spectroscopic anomaly: the Zeeman effect. In 1896 Pieter Zeeman observed that when a sodium flame was placed between the poles of a Ruhmkorff electromagnet the two lines of the first principal doublet were considerably broadened.[28] One year later Lorentz provided a classical explanation for this phenomenon.[29] The external magnetic field **H** was supposed to induce a change in the motion of the electrons. The electrons, whose orbit-planes were normal to the direction of **H**, were speeded up (if moving in a counter-clockwise direction) or slowed down (if moving in a clockwise direction) from their usual orbital frequency $\nu_0$ by an amount of respectively $\pm\Delta\nu$ depending on the strength of the field, the charge-to-mass ratio of the electron and the velocity of light (i.e. $\Delta\nu = eH/(4\pi mc)$). It was soon realised that $\Delta\nu$ was nothing but Larmor's frequency. If the electrons' motions are viewed parallel to the direction of the field, the emitted light is right- and left-handed circularly polarized so that a doublet is observed (Fig. 2.2B) with frequency $\nu_d = \nu_0 \pm \nu_L$. On the other hand, if viewed perpendicular to the field, three components are observed: a central unshifted line and two lines displaced on either side of the central one, with

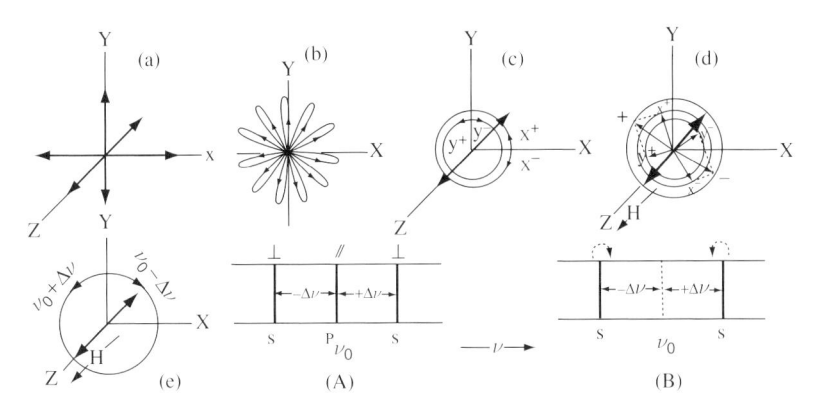

Fig. 2.2 Diagrams for Lorentz's explanation of the normal Zeeman effect, where (A) is the diagram for the triplet and (B) for the doublet.

*Source*: H. E. White (1934) *Introduction to Atomic Spectra* (New York, London: McGraw-Hill Book Company). Reproduced with permission of The McGraw-Hill Companies.

---

[28] Zeeman (1896).  [29] Lorentz (1897).

frequency, $\nu_t = \nu_0, \nu_0 \pm \nu_L$ respectively (Fig. 2.2A). This triplet is called the *Lorentz normal triplet* and is the spectral pattern of the so-called *normal Zeeman effect* typical of the series spectra of zinc, copper, and cadmium, among others.

The low resolving power of the instruments did not actually allow Zeeman to observe doublets and triplets; he could only observe that the lines were widened in the presence of a magnetic field. It was Preston who first observed – using instruments with greater resolving power – that not only were certain lines of the zinc spectrum split up into triplets when viewed perpendicular to the field, but also that others were split into as many as four and even six components. In December 1897, Preston reported to the Royal Dublin Society that the two D lines of sodium were split into a quadruplet and a sextuplet (Fig. 2.3).

Similar anomalous patterns were later observed in the series spectra of other elements and they all fell under the name of *anomalous Zeeman effect*. A few years later, Runge[30] pointed out that the frequencies of the components of the anomalous Zeeman pattern were rational fractions – called Runge fractions – of the so-called Lorentz unit of the Zeeman effect.[31] Despite Runge's law and similar empirical rules, an explanation of the anomalous Zeeman effect was beyond the reach of spectroscopy until 1921, when Landé made an important contribution. He managed to

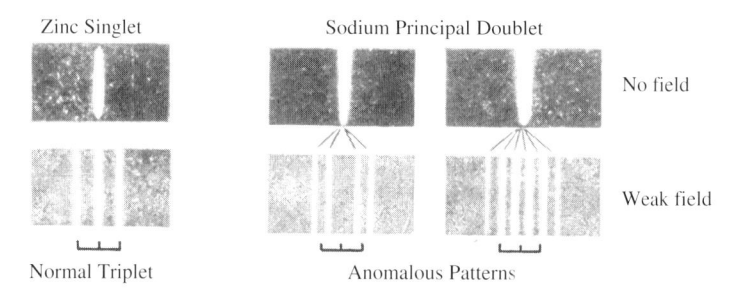

Fig. 2.3 Spectral patterns of the normal Zeeman effect of zinc and of the anomalous Zeeman effect of sodium, viewed perpendicular to the magnetic field.

*Source*: H. E. White (1934) *Introduction to Atomic Spectra* (New York, London: McGraw-Hill Book Company). Reproduced with permission of The McGraw-Hill Companies.

---

[30] Runge (1907).
[31] For instance, the anomalous Zeeman pattern of the sodium principal series doublet can be expressed as $\pm 2/3\mathbf{L}$ for the quadruplet and $\pm 4/3\mathbf{L}$ for the sextuplet, where $\mathbf{L}$ is the Lorentz unit of the Zeeman effect $\mathbf{L} = (eH/(4\pi mc^2))$.

## 2.1 The prehistory of Pauli's exclusion principle

derive the anomalous doublets, but at the cost of strengthening the doublet riddle and the riddle of statistical weights.

Landé[32] empirically ascribed half-integral values to the magnetic quantum number (and hence also to the inner quantum number $j$): $m = \pm 1/2$, $\pm 3/2$, $\pm 5/2$, ..., $\pm(j-1/2)$ so that he could derive doublets. But half-integral values for $j$ and $m$ clearly violated the Bohr–Sommerfeld theory, which allowed only integral values for quantum numbers, as well as Bohr's building-up principle, which forbade the decrease of a half-integral unit in the building-up process. The riddle of statistical weights was still looming on the horizon. Furthermore, half-integral quantum numbers for alkali doublets were in contrast with relativistic X-ray doublets that required only integral quantum numbers: Landé's proposal strengthened the doublet riddle.

Landé also calculated the energy $W$ of a Zeeman state (i.e. the Zeeman Hamiltonian describing the interaction energy between the external magnetic field and the magnetic moment of the atom) as

$$W = W_0 + [mgh(eH/4\pi mc)] = W_0 + mgh\nu_L \qquad (2.11)$$

where $W_0$ is the Hamiltonian of the atom in the absence of an external field, $m$ is the magnetic quantum number,[33] $\nu_L$ is the Larmor frequency and $g$ is the so-called *Landé g factor*. The $g$ factors were introduced on a purely empirical basis as *Grundenergieniveau* specific for each spectral term.[34] The magnetic quantum number $m$ times $g$ turned out to give the right splitting factors for the anomalous doublets. For instance, by empirically fixing the $g$ values for the following spectral terms of the sodium principal and sharp series doublets,[35] the resultant $mg$ splitting factors could be obtained:

$$g = 2/3 \quad \text{for the term } {}^2P_{1/2} \implies mg = \pm 1/3$$

$$g = 4/3 \quad \text{for the term } {}^2P_{3/2} \implies mg = \pm 2/3, \pm 6/3$$

$$g = 2 \quad \text{for the term } {}^2S_{1/2} \implies mg = \pm 1$$

---

[32] Landé (1921a).
[33] Notice the difference between the first $m$ in the formula (the magnetic quantum number), and the second $m$ in the denominator, which denotes the electron's mass in the charge-to-mass ratio of the Larmor frequency.
[34] Landé (1921a). For a historical reconstruction, see Forman (1970).
[35] Here the subscript of the spectral term denotes Landé's half-integral $j$ as well as the corresponding magnetic number $m$, e.g. the term ${}^2P_{1/2}$ corresponds to $j = 1/2$ and $m = \pm 1/2$. For details see White (1934), p. 158.

Landé proceeded to generalize this result to singlets and triplets. But how to justify the *g* factors and their fractional values? How to explain the $2(2k-1)$ states observed in the presence of a magnetic field? Landé's phenomenological analysis left these questions unanswered. As we shall see in the following sections, it was Heisenberg's, Bohr's, and Pauli's task to address them and find possible solutions.

Two years later, Landé returned to the problem.[36] By using the atomic core model he found an empirical formula for the *g* factors, which turned out to depend on the quantum numbers *j*, *r* and *k*:

$$g = \frac{3}{2} - \frac{1}{2}\frac{\left(k^2 - r^2\right)}{\left(j^2 - \frac{1}{4}\right)} \tag{2.12}$$

Setting $k = l + 1/2$, $r = s + 1/2$, and $j = j' + 1/2$, Landé's formula corresponds to the current formula for the *g* factor, in modern notation

$$g = \frac{3}{2} - \frac{1}{2}\frac{l(l+1) - s(s+1)}{j'(j'+1)} \tag{2.13}$$

where the spin quantum number *s* has replaced the core quantum number *r* and $j' = \sqrt{j(j+1)}$. Landé's procedure was entirely a posteriori: both the *g* factors and the formula (2.12) were obtained from a phenomenological analysis of spectra. Looking for a justification, Landé hypothesized – in the light of the atomic core model – that the *mg* splitting factor corresponded to the sum of the projections of the electron angular momentum vector **K** and the core angular momentum vector **R** on the external field vector **H**. Both **K** and **R** Larmor-precessed around the resultant total angular momentum vector **J**, with **J** in turn Larmor-precessing around **H** with angular velocity $g\omega_L$. There was however a puzzling anomaly concerning the observed unexpected value 2 of the *g* factor. In other words, the *mg* splitting factor was equal to

$$mg = |\mathbf{K}|\cos(\mathbf{K},\mathbf{H}) + 2|\mathbf{R}|\cos(\mathbf{R},\mathbf{H}) \tag{2.14}$$

rather than

$$mg = |\mathbf{K}|\cos(\mathbf{K},\mathbf{H}) + |\mathbf{R}|\cos(\mathbf{R},\mathbf{H}) \tag{2.15}$$

[36] Landé (1923a).

## 2.1 The prehistory of Pauli's exclusion principle

as was expected from Larmor's theorem applied to **K** and **R**. This suggested that **R** precessed twice as fast around **H** as **K** did, i.e. **R**'s precessional angular velocity seemed to be twice Larmor's angular velocity $\omega_L$. This faster precession was ascribed to an anomalous magneto-mechanical ratio between the core magnetic moment $\mu_r$ and the core angular momentum $p_r$, a ratio that was supposed to be twice the corresponding magneto-mechanical ratio for the electron

$$\mu_r/p_r = 2(\mu_k/p_k) = 2(e/2mc) \qquad (2.16)$$

There were only two alternatives to account for this 'core magnetic anomaly', experimentally detected a few years earlier:[37] either to modify Larmor's theorem, or to postulate a further rotation of the core so as to account for its faster precessional angular velocity. Landé first tried to modify Larmor's theorem without success. In 1923 he was driven to the second alternative.[38] Both alternatives were wrong and a conclusive understanding of this anomaly came only with Pauli's rejection of the atomic core model and the later introduction of the electron spin, whose magneto-mechanical ratio is exactly twice the orbital magneto-mechanical ratio. Furthermore, as we shall see in Chapter 4, only with Dirac's equation for the electron in 1928 did it become possible to derive the spin magnetic moment, and to explain the anomalous value 2 of the $g$ factor. This was indeed one of the most important achievements of the new quantum theory. But in the framework of the (pre-spin) old quantum theory, the culprit to blame for the observed anomaly was the atomic core.

Landé's analysis of the anomalous Zeeman effect left some important questions unanswered. First, the failure of Larmor's theorem in the anomalous Zeeman effect needed to be explained. Second, half-integral values cried out for a theoretical explanation: they were postulated to accommodate spectroscopic data, although they were not allowed in the Bohr–Sommerfeld theory (doublet riddle and riddle of statistical weights). Should this theoretical impasse be overcome by abandoning the Bohr–Sommerfeld theory together with the building-up principle? Or should half-integral quantum numbers be rejected as non-orthodox? Could the two horns of this dilemma be reconciled? In the short but intense

---

[37] In 1919 E. Beck performed a series of experiments that shed light on the existence of an anomalous magneto-mechanical ratio for all ferromagnetic materials. However, Beck thought that such an anomalous ratio was due to the electron's orbital motion, and not to the atomic core as later Landé (and after him Heisenberg) supposed.

[38] An attempt to modify Larmor's theorem without any compelling theoretical reason is given in Landé (1921b), while the second option is discussed in Landé (1923a).

period between 1921 and 1924 Bohr, Heisenberg, and Pauli addressed these questions and proposed alternative solutions that I now turn to in some detail.

## 2.2 Bohr, Heisenberg, and Pauli on spectroscopic anomalies

### 2.2.1 Niels Bohr: nothing but a 'non-mechanical constraint'?

In an article written for a special issue of *Annalen der Physik* dedicated to the spectroscopist Kayser, Bohr offered a survey of the state of the art in spectroscopy up to 1923.[39] Starting with the quantum theory of conditionally periodic systems and the correspondence principle, Bohr investigated how this theoretical framework fared on the score of spectroscopic findings. While the treatment of one-electron atom spectra was fairly unproblematic, an adequate treatment of many-electron atom spectra had proved more complicated. Even more serious difficulties arose in the case of the anomalous Zeeman effect and the Paschen–Back effect:[40] while the former concerns spectral patterns in a weak magnetic field, the latter concerns spectral patterns in a strong magnetic field.[41] Bohr could not help noting that Landé's experimental findings about the anomalous Zeeman effect were at odds with the quantum theory of periodic systems:

> It follows from Landé's analysis that the value of the component of the angular momentum of the atom parallel to the field, divided by $2\pi$, may not always be equal to an integer, as assumed in the interpretation of the Zeeman effect of the hydrogen lines ... Landé actually obtained values for this quantity ... differing from an integer by 1/2. This finding has made questionable the justification of fixing the motion in the stationary states of an atom with several electrons in direct analogy with the quantum theory of periodic systems.[42]

Not disheartened by Landé's experimental findings, Bohr put forward a very speculative hypothesis that could accommodate evidence. Following the atomic core model, Bohr assumed that the spectral multiplicity arose

---

[39] Bohr (1923a).    [40] Paschen and Back (1921).
[41] In its simplest form, the Paschen–Back effect concerns the appearance of triplets that resemble the Lorentz normal triplets of the normal Zeeman effect. In the terminology of the semi-classical vector model, as the doublet splitting in the anomalous Zeeman effect is due to the spin–orbit coupling, in which both $k$ and $s$ precess (with $s$ twice as fast as $k$), analogously the triplets of the Paschen–Back effect are due to the spin–orbit *de*coupling: in a strong magnetic field, $k$ and $s$ get decoupled, separately quantized and precess around the field independently of each other.
[42] Bohr (1923a). Reprinted in Bohr (1977), p. 630.

## 2.2 Bohr, Heisenberg, and Pauli on spectroscopic anomalies

from the coupling of the core's inner orbits with the valence electron's outer orbit. This resulted in radiative transitions not corresponding to the fundamental harmonic component of the electronic motion, but to smaller harmonic components: the correspondence principle still offered the general framework for an understanding of complex spectra. Furthermore, he noticed a 'recurring closure of certain electron groups with increasing atomic number, a closure expressing itself in a disinclination of these groups to admit any additional electrons in orbits with the same values of the quantum number $n$ and $k$'.[43] As we shall see in Section 2.2.4, what Bohr here mildly describes as a 'disinclination' of groups to admit additional electrons in orbits with the same quantum numbers soon turned out to be a 'prohibition' according to Pauli's exclusion rule. By contrast, Bohr traced this 'disinclination' back to the symmetry requirements he envisaged for the closure of electronic orbits of noble gases,[44] even if, as mentioned in Section 2.1.1, these symmetry requirements implied a problematic reopening of already closed electronic orbits.

The interplay between inner and outer orbits (i.e. the core–electron coupling) could – once suitably adjusted – account both for the alkali doublets and for the anomalous Zeeman effect. In the case of alkalis, Bohr fixed the $j$-values of the doublet p-terms ($k = 2$) equal to $k + d$ and $k + d - 1$, whereas for the singlet s-term ($k = 1$) $j = 1 + d$, 'where $d$ is a quantity that is independent of $k$, but is to begin with undetermined'.[45] Bohr's explanation of the anomalous Zeeman effect was even more ingenious. As Landé's analysis had revealed, an atom placed in a weak magnetic field could take on $2(2k - 1)$ states. But this number was not equal to the product of the number of electron states with the number of core states in the magnetic field. In fact, the atomic core could take on only one state, i.e. perpendicular to the field direction, by analogy with noble gases. On the other hand, the valence electron could take on only $2k$ positions in the field so that the product of the respective positions of the core and electron was $2k$, a number half as large as Landé's experimentally detected $2(2k - 1)$. To accommodate this discrepancy, Bohr advanced the following suggestion:

---

[43] *Ibid.*, p. 642.
[44] 'We may assume that, in the normal states of the atoms of the noble gases, we are dealing with electron configurations of pronounced symmetry. Actually, we see just in this symmetry the explanation of the closed character of these electron groups; for the admission of additional electrons, which would destroy this symmetry, would, in fact, not show that analogy with a classical radiation process which could be established for the possible transitions between stationary states of simply or multiply periodic systems.' *Ibid.*, p. 645.
[45] *Ibid.*, p. 645.

> We are led to the view that, because of stability properties of the atom which cannot be described mechanically, the coupling of the series electron to the atomic core is subject to a constraint (*Zwang*) which is not analogous to the effect of an external field, but which forces the atomic core to assume two different positions in the atom, instead of the single orientation possible in a constant external field, while at the same time, as a result of the same constraint, the outer electron, instead of $2k$ possible orientations in an external field, can only assume $2k-1$ orientations in the atomic assemblage.[46]

A 'non-mechanical constraint'[47] (*unmechanischer Zwang*) not analogous to the action of any external field was supposed to intervene in the electron–core coupling to set right the total number of the states in the anomalous Zeeman effect: the *Zwang* would subtract one position from the $2k$ positions of the electron (which then remained $2k-1$) and would add it to the core that as a result would take two possible positions, instead of one.[48] As if by magic, the total number of anomalous Zeeman states turned out to be $2(2k-1)$. The 'non-mechanical constraint' was plucked out of the air as an ad hoc auxiliary hypothesis to retain the validity of the atomic core model, and to reconcile it with recalcitrant evidence. But its physical nature remained obscure: since it was not analogous to the effect of a magnetic or electric field, what kind of physical constraint was it and how did it originate? Without answering these questions, Bohr contented himself with reconciling theory and evidence.

Bohr was however well aware that this way of amending the atomic core model was in contrast with the quantization rules for periodic systems. A clear sign of this contrast was the half-integral values of the magnetic quantum number $m$ and inner quantum number $j$ introduced by Landé:

> This finding makes it natural to surmise that the orientation of the orbit of the series electron relative to the atomic core cannot be described either by integer values of $j$; in particular, for reasons of symmetry, the formal assumption $d = 1/2$ might offer itself.[49]

Given the total angular momentum $j$ of the atom, saying that its half-integral values are due to the sum of $k$ with a new quantity $d = 1/2$, is empirically equivalent to saying, in the later semi-classical terminology, that the total

---

[46] *Ibid.*, p. 646.
[47] The adjective 'non-mechanical' was later added to Bohr's 'constraint' by Heisenberg (1925a). On Bohr's *unmechanischer Zwang*, see Serwer (1977).
[48] The hypothesis of a non-mechanical force not analogous to a Coulomb force had been previously used by Van Vleck (1922) to explain the anomalous instability of the helium model.
[49] Bohr (1923a). Reprinted in Bohr (1977), p. 647.

angular momentum of the atom, which is contributed by the valence electron alone, is made up of two parts: the electron's orbital angular momentum (given by $k$) and the electron's spin angular momentum ($s = \pm 1/2$) so that $j = k \pm s$, hence the doublet p-terms $j = k + 1/2$ and $j = k - 1/2$. Bohr's quantity $d$ played a role empirically equivalent to that of the electron's spin $s$ to bring about the half-integral values of the total angular momentum, and hence the doublet terms. However, $d$ did not anticipate the electron's spin, because it was not meant to be a property of the electron. It was rather an empirically fixed quantity suitably chosen to accommodate experimental evidence, but whose physical meaning was as obscure as that of the *Zwang*.

Surprisingly enough, half-integral values for $j$ and $m$ did not undermine the validity of the building-up principle, whose scope was restricted to quantum numbers for individual electrons in $n_k$-orbits. Thus, while allowing half-integral values for $j$ and $m$, Bohr insisted that half-integral values for the azimuthal quantum number $k$ 'must be regarded as a departure from [the quantum theory of periodic systems] which can hardly be substantiated', and whose consequences 'seem to contradict our experience about spectra'.[50] The target of this blow, explicitly mentioned in the same passage, was a bold version of the atomic core model elaborated by Werner Heisenberg in 1921 that relied precisely on unorthodox (building-up principle violating) half-integral $k$ and to which I now turn.

### 2.2.2 Heisenberg's first core model: the sharing principle. Does success justify the means?

After Landé's empirical findings on the anomalous Zeeman effect,[51] Heisenberg too tried to accommodate the anomalous evidence with the atomic core model. To this purpose he elaborated a first version of the model in 1922, and a second improved one in 1924, known in the historical literature respectively as Heisenberg's first and second core models. The first core model[52] hinged on the so-called *sharing principle*: the valence electron was supposed to 'share' half of its angular momentum with the core, which was taken as originally having no net angular momentum. The quantum condition for the electron's angular momentum was fixed as

---

[50] *Ibid.*, footnote on p. 647.  [51] Landé (1921a), (1921b).
[52] Heisenberg (1922). For a detailed historical analysis of the first core model, see Cassidy (1979) on which I draw for this section.

$$\int_0^{2\pi} p \, d\beta = k^* h \qquad (2.17)$$

where $p$ is the electron's angular momentum, $\beta$ is the azimuthal angle, and $k^*$ is the azimuthal quantum number that under the *sharing principle* was set equal to $k - 1/2$ (the asterisk indicates that the number was half-integral valued). While Landé and Bohr retained integral values for $k$ as required by the Bohr–Sommerfeld quantum conditions (2.5)–(2.6), Heisenberg assigned it – in a completely unorthodox way – half-integral values ($k = 1/2, 3/2, \ldots$). On the other hand, the core picked up the half-quantum from the valence electron so that its angular momentum was equal to $1/2$. In this case, it was the core (not Bohr's quantity $d$) that played a role empirically equivalent to that of the electron spin to bring about the alkali doublets and their Zeeman splitting. The half-integral values for $k^*$ and $r^*$ allowed Heisenberg to obtain doublets in agreement with Sommerfeld, who for $k = 2$ (i.e. p-terms) had fixed the doublet states $j = k$ and $j = k - 1$. Summing $k^*$ and $r^*$ Heisenberg got respectively $j = (k - 1/2) + 1/2 = k$ and $j = (k - 1/2) - 1/2 = k - 1$.

In the vectorial notation, the electron's angular momentum vector **k**\* was supposed to move around the core's angular momentum vector **r**\* with an angle $\beta$ in such a way that an internal magnetic field $\mathbf{H}_{Int}$ (whose direction coincided with **k**\*) was created. The core precessed about $\mathbf{H}_{Int}$ with interaction energy

$$\Delta E = 1/2 \cdot h/2\pi \cdot e/2mc \cdot |\mathbf{H}_{Int}| \cos \beta = 1/2 h\nu_L \cos \beta \qquad (2.18)$$

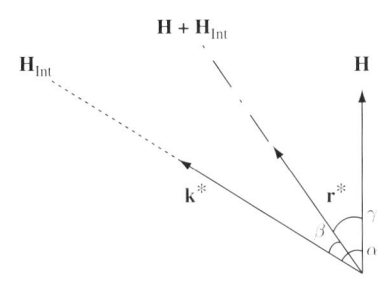

Fig. 2.4 The core–electron coupling in the presence of an external magnetic field **H**, according to Heisenberg's first core model. Adapted from Cassidy (1979).

## 2.2 Bohr, Heisenberg, and Pauli on spectroscopic anomalies

where $\nu_L$ is the Larmor frequency. In the presence of an external magnetic field **H**, because of the interaction between the internal $\mathbf{H}_{Int}$ and the external field **H**, the core was supposed to stay along the resultant $\mathbf{H} + \mathbf{H}_{Int}$ (Fig. 2.4) with corresponding interaction energy

$$\Delta E = h\nu_L (k^* \cos\alpha + r^* \cos\gamma + \nu \cos\beta) \quad (2.19)$$

where $\nu = |\mathbf{H}_{Int}|/|\mathbf{H}|$; $k^* \cos\alpha$ was the projection of the electron angular momentum on **H**;[53] $r^* \cos\gamma$ was the projection of the core angular momentum ($r^* = 1/2$) on **H**; and $\nu \cos\beta$ denoted the internal magnetic interaction between the core and the electron. Heisenberg gave an explanation of the anomalous Zeeman effect as well as of the Paschen–Back effect using simple geometrico-mechanical properties of this model. He pointed out that if **H** increases (strong magnetic field), the angle $\beta$ between $\mathbf{H}_{Int} + \mathbf{H}$ and $\mathbf{H}_{Int}$ increases. Accordingly, the ratio $\nu = |\mathbf{H}_{Int}|/|\mathbf{H}|$ decreases, with the consequence that the internal field $\mathbf{H}_{Int}$ is gradually overcome and the decoupling of the two fields causes the appearance of the triplet typical of the Paschen–Back effect.

On the other hand, if **H** is small (weak magnetic field), the ratio $\nu$ increases ($\nu \to \infty$), while $\beta$ decreases so that $\cos\beta$ approaches $\pm 1$. If $\cos\beta = -1$, the core (which, recall, was supposed to stay along the $\mathbf{H}_{Int} + \mathbf{H}$ axis) is aligned parallel to $\mathbf{H}_{Int}$. If $\cos\beta = +1$, the core is aligned anti-parallel to $\mathbf{H}_{Int}$. This simple geometrical picture accounted for the doublet fine structure. The parallel or anti-parallel alignment of the core vector $r^*$ with the electron vector $k^*$ yielded a result empirically equivalent to the one obtained by taking into account the spin–orbit coupling contribution in the fine structure Hamiltonian. In other words, the total angular momentum $j = k^* \pm r^*$ was empirically equivalent to what in the later semi-classical spinning electron model would be $j = l \pm s$, with the spin magnetic moment $\mu_s$ aligned either anti-parallel or parallel to **H**, respectively (see Fig. 2.5). From this model, Heisenberg could derive the Sommerfeld–Voigt formula, from which the Landé g factors for doublets could be retrieved.[54] The calculated splitting for the lithium 2p level,

---

[53] In contrast with Landé, who identified the magnetic quantum number $m$ with the projection of the total angular momentum on **H**, Heisenberg identified $m$ with the projection of the electron angular momentum. Given the half-integral values for $k^*$, $m^*$ also took half-integral values ($m^* = \pm 1/2, \pm 3/2, \ldots, \pm k^*$) in agreement with Landé.

[54] In September 1921 Sommerfeld (1922) was working on the Voigt theory of the anomalous Zeeman effect for sodium D lines. Only with Heisenberg's introduction of half-integral $k$ could the Sommerfeld–Voigt formula yield the optical doublets. See Cassidy (1979), pp. 196–202.

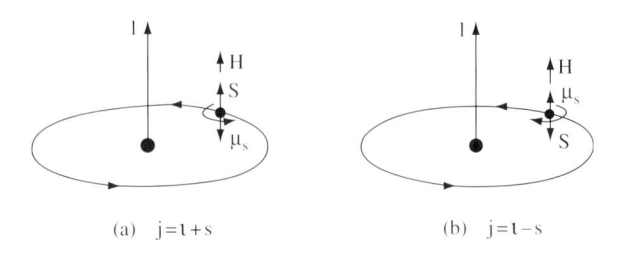

Fig. 2.5 The spin–orbit coupling for the two fine-structure states (a) $j = l + 1/2$ and (b) $j = l - 1/2$, according to the semi-classical spinning electron model.

*Source*: H. E. White (1934) *Introduction to Atomic Spectra* (New York, London: McGraw-Hill Book Company). Reproduced with permission of The McGraw-Hill Companies.

$\Delta\nu_{2p} = 0.32 \, \text{cm}^{-1}$, was close to the experimentally found value, $\Delta\nu_{2p} = 0.34 \, \text{cm}^{-1}$.

Despite the relative success in recovering both the Paschen–Back effect and the anomalous Zeeman effect, the model faced serious theoretical difficulties in the case of intermediate fields. As mentioned, Heisenberg assumed that the core stayed along the resultant axis $\mathbf{H}_{\text{Int}} + \mathbf{H}$, and that the transition from weak to strong fields was an adiabatic process. Thus, for intermediate fields, the core was forced to stay invariantly in the direction of $\mathbf{H}_{\text{Int}} + \mathbf{H}$, without precessing about the electron vector $\mathbf{k}^*$ under the effect of the internal magnetic field $\mathbf{H}_{\text{Int}}$. But this implied (1) a violation of Larmor's theorem; (2) a violation of Rubinowicz's integral selection rules $\Delta j = 0, \pm 1$ and $\Delta m = 0, \pm 1$.

The violation of Larmor's theorem was not a novelty. As we saw in Section 2.1.3, Landé's analysis had already revealed that Larmor's theorem was violated in the anomalous Zeeman effect, and he even suggested modifying it. Pauli too – as we shall see later in this chapter – had to deal with the failure of Larmor's theorem, and indeed this was the main stumbling-block towards his final breakthrough. But it is one thing to claim that Larmor's theorem fails because of a core magnetic anomaly (as Landé mistakenly supposed), or because of the anomalous spin magnetic moment (as it was realized after Pauli's breakthrough). It is another thing to claim, as Heisenberg did, that Larmor's theorem applies both to the core–field and to the electron–field interaction, but not to the core–electron interaction. Given the internal magnetic field $\mathbf{H}_{\text{Int}}$, why should not the core precess about it? This was a completely arbitrary and unjustified deviation from Larmor's theorem.

Even worse was the violation of Rubinowicz's selection rules, which Landé complained of to Heisenberg. If the core is forced to stay invariantly

## 2.2 Bohr, Heisenberg, and Pauli on spectroscopic anomalies 59

along the $\mathbf{H}_{Int} + \mathbf{H}$ direction, it can take on any continuous value. This violated space quantization as well as the Rubinowicz rules: the total angular momentum was not restricted to a set of discrete values. And since Rubinowicz's rules were obtained from the conservation of the angular momentum in the classical theory of radiation, their violation could be interpreted as due either to a violation of the conservation of angular momentum or to a violation of the classical theory of radiation. Between the two, Heisenberg sacrificed the latter by introducing non-classical waves with fractional angular momenta, whose sum supposedly yielded a classical wave with integral angular momentum.

This solution could not however prevent Heisenberg's model from facing even more theoretical difficulties, namely those Bohr was concerned with. Fractional angular momenta violated the Bohr–Sommerfeld quantum conditions. Furthermore, half-integral $k$ were also a departure from the relativistic X-ray doublets, for which the usual integral values were used. Thus, Heisenberg's first core model accounted for the optical doublets in a way that contradicted the treatment of relativistic doublets: the theoretical unification of optical and relativistic doublets was still far away and the doublet riddle unsolved.

No less problematic was the half-integral value of the core. By picking up half-quantum angular momentum from the valence electron, the atomic core patently violated Bohr's building-up principle. The quantum numbers and the statistical weights did not remain adiabatically invariant. The sharing principle strengthened the riddle of statistical weights.

To sum up, Heisenberg's first core model could empirically entail alkali doublets and their Zeeman splitting. But the price to pay for this empirical success was the violation of several theoretical assumptions: Larmor's theorem, Rubinowicz's selection rules, Sommerfeld's space quantization, the classical theory of radiation, and the Bohr–Sommerfeld quantum conditions. Heisenberg announced his result in a letter to Pauli. Intellectual honesty compelled him to admit the 'drawbacks of this theory'. Yet he optimistically concluded with the Machiavellian motto that 'despite all: success justifies the means'.[55] But, as a matter of fact, empirical success does *not* justify the use of *whatever* means. No wonder then that after two years Heisenberg proposed a second improved core model, which in his expectations would succeed where the first model failed.

---

[55] Heisenberg to Pauli, 19 November 1921. Pauli (1979), p. 44.

### 2.2.3 Heisenberg's second core model: the branching rule and a new quantum principle

Heisenberg's first core model gave rise to a lively discussion. The theoretical scenario was messy enough to urge 'the introduction of new hypotheses – either new quantum conditions or proposals for changing mechanics'.[56] The year 1923 was the *annus mirabilis* for the history of the anomalous Zeeman effect. Landé[57] published his result on the g formula (2.12) and the core magnetic anomaly, as we saw in Section 2.1.3. In September, Pauli wrote to Landé that the quantities figuring in Landé's formula for the g factors (namely $r$, $k$ and $j$) were not to be taken as the true [*wahren*] angular momenta of the core, the valence electron, and the atom, respectively. The 'true' angular momenta were rather given by $R = r + 1/2$, $K = k + 1/2$, $J = j + 1/2$:

> It seems that each angular momentum is not represented through one quantum number, but rather through a pair of numbers. Under certain respects, angular momenta appear to be twofold [*zweideutig*]. For instance, the invariance of the statistical weights is violated during the [electron–core] coupling because the core angular momentum is given by either of the numbers. Notice moreover that this 'twofoldness' [*Zweideutigkeit*] concerns also $k$. According to this point of view, there is no half-integral $k$. This seems to be in a better agreement also with the X-rays.[58]

In this historically important passage, Pauli anticipated an idea that turned out to be crucial for later developments: the *Zweideutigkeit* of the angular momentum. By this vague expression Pauli denoted a no better qualified 'twofoldness' – so to speak – of the angular momentum. A year later Pauli came to use this expression to refer exclusively to the electron's angular momentum, as I shall discuss in detail in the next section. The historically important fact is that already in September 1923 Pauli had introduced the notion of *Zweideutigkeit* to denote a 'twofold' angular momentum. Nor did it refer to the electron's angular momentum in particular, but to *any angular momentum*, i.e. of the core, the electron as well as of the atom. Let me then underline two significant consequences of this theoretical step:

(i) The *Zweideutigkeit* of the core's angular momentum shed light on the riddle of statistical weights: the core of an alkali ion did not coincide

---

[56] Heisenberg to Pauli, 19 February 1923. In Pauli (1979), p. 80.   [57] Landé (1923a).
[58] Pauli to Landé, 23 September 1923. In Pauli (1979), p. 123.

with the preceding noble gas because there were now two possibilities for its angular momentum, either $r$ or $r + 1/2$.

(ii) The *Zweideutigkeit* of the electron's angular momentum shed light on the problematic half-integral values for $k$, which turned out to be due to $k + 1/2$, while integral-valued $k$ was in agreement with X-ray spectra.

Pauli's notion of *Zweideutigkeit* exerted a direct influence on Heisenberg. In May 1924, Heisenberg and Landé published a joint article on the term-structure of higher-level multiplets,[59] starting from Landé's recent analysis of the neon spectrum.[60] The neon ion consisted of two p-terms: a $p_1$-term with $j = 2$ and a $p_2$-term with $j = 1$. The addition of the outer electron in the building-up process apparently made the $p_1$-term split into a triplet and a quintuplet, while the $p_2$-term split into a singlet and a triplet. Hence:

(i) from the neon ion ($j = 1$) with the addition of the valence electron, two s-terms were derived with respectively $j = 1/2$ and $j = 3/2$;
(ii) from the neon ion ($j = 2$) another two s-terms were derived with $j = 3/2$ and $j = 5/2$.

Both these results seemed to spring from $j$ 'branching' [*Verzweigung*] into $j \pm 1/2$ with the addition of the valence electron.[61] Landé and Heisenberg generalized this result in a 'branching rule' [*Verzweigungsregel*]:

> When the atom is built up from the states of the ion that are characterized through the $j$-values $j_1, j_2, \ldots, j_n$ we assume that the atom possesses $2n$ s-terms with the $j$-values $j_1 \pm 1/2, \ldots, j_n \pm 1/2$, and accordingly the atom shows $2n$ multiplet-systems with multiplicity $2j = 2(j_1 \pm 1/2), 2(j_2 \pm 1/2), \ldots$ and so on.[62]

This new rule, phenomenologically derived from the neon spectrum, could account for the multiplet structure. Pauli's *Zweideutigkeit* was tacitly at work in the Heisenberg–Landé branching rule. The *Zahlenpaar* that Pauli regarded as the 'true' angular momenta had here become the 'branching' quantum numbers of the total angular momentum. Yet the rule violated Bohr's building-up principle. Given the 'branching' process, the terms of the atom could not be expected to coincide with those of the ion. The riddle of statistical weights, once again, loomed on the horizon. Landé and Heisenberg could not help concluding the article with an apologetic remark:

---

[59] Landé and Heisenberg (1924).  [60] Landé (1923c).
[61] Landé and Heisenberg (1924), p. 280.  [62] *Ibid.*, p. 284.

> This formally so easily representable branching process of the angular momentum has become one of the main objections against the applicability of the quantum rule, valid for conditionally periodic motions, to coupled systems and it has consequently led to search for a simple modification of the quantum rules used so far.[63]

A footnote accompanied this last sentence mentioning a forthcoming work of Heisenberg in *Zeitschrift für Physik*. One month later (in June 1924) Heisenberg published an article significantly entitled 'On a modification of the formal rules of quantum theory in the problem of the anomalous Zeeman effect'.[64] Apropos of this, the historian Daniel Serwer has noticed that the branching rule was a sort of publicity gimmick to smooth the path to the introduction of Heisenberg's new quantum principle, which solved both the riddle of statistical weights and the doublet riddle.[65] But I think that the branching rule smoothed the path to Heisenberg's new quantum principle precisely because it was, in Heisenberg's words, 'directly equivalent' to it.[66] Indeed, the new quantum principle was nothing but the Heisenberg–Landé branching rule (inspired by Pauli's *Zweideutigkeit*) but this time devoid of the problematic branching process.

Pauli received Heisenberg's draft in December 1923 together with a letter in which Heisenberg asked him '1. if you regard it as totally rubbish, 2. and if this is not the case, (a) would you be so kind to send it to Bohr for criticisms ... since I would like to have the "papal blessing" before the publication'.[67] As a reply, Bohr invited Heisenberg to Copenhagen in spring 1924, where under Bohr's influence Heisenberg modified the draft, finally published in June 1924.[68] At the beginning of the article, Heisenberg recalled the riddle of statistical weights: $j$ was expected to be invariant according to the building-up principle, and yet spectroscopic evidence suggested $j + 1/2$ and $j - 1/2$. Hence the necessity to introduce a new quantum rule:

> A determinate value of the coupling energy between the electron and the atomic core is not associated, as it has been so far assumed, to one value of the inner quantum number $j$, but rather with two values ($j + 1/2$ and $j - 1/2$ in our numeration).[69]

---

[63] *Ibid.*, p. 286.   [64] Heisenberg (1924).   [65] See Serwer (1977), p. 217.
[66] Heisenberg (1924), p. 301.
[67] Heisenberg to Pauli, 7 December 1923. In Pauli (1979), p. 132.
[68] The original title of Heisenberg's work was 'Über ein neues Quantenprinzip und dessen Anwendung auf die Theorie der anomalen Zeemaneffekte'; a copy of the original manuscript is in AHQP (45, 8).
[69] Heisenberg (1924), p. 292.

## 2.2 Bohr, Heisenberg, and Pauli on spectroscopic anomalies

As Heisenberg himself admitted, this was nothing but the Heisenberg–Landé branching rule suitably devoid of the branching process. Starting from it, Heisenberg formulated a new quantum principle. The coupling energy between the core and the electron was given by

$$H_{qu} = \frac{\partial F}{\partial j} \qquad (2.20)$$

By plugging in the derivative $\Delta F = F(j + 1/2) - F(j - 1/2)$, the quantum mechanical Hamiltonian $H_{qu}$ turned out to be

$$H_{qu} = \int_{j-1/2}^{j+1/2} H_{cl}\, dj \qquad (2.21)$$

This formula suggested a correspondence between the quantum mechanical Hamiltonian $H_{qu}$ and the classical Hamiltonian $H_{cl}$. Not only was $H_{qu}$ obtained by integrating over $H_{cl}$ in the interval imposed by the new quantum rule, i.e. from $j - 1/2$ to $j + 1/2$, but 'in the case of high quantum numbers, they almost coincide'.[70] There could hardly be a better choice to get Bohr's 'papal blessing' than elaborating a new quantum principle along the lines of Bohr's correspondence principle.

Implementing the above formula in the *Ersatzmodell*,[71] Heisenberg obtained the Landé $g$ factors, under Pauli's proviso that *zweideutig k*, $r$ and $j$ replaced Landé's integral quantum numbers. But what was the physical meaning of Heisenberg's new quantum principle?

> The more the rule ... simplifies the problem of the Zeeman effect in the formal respect, the more dubious its physical interpretation seems. *Prima facie* it seems an unusual feature of the theory that a value of the coupling energy with two quantum numbers $j$ is associated ... But perhaps the formal analogy between [the new principle] and the frequency condition could later provide a physical interpretation of [the new principle], because the problem of the coupling of electrons is closely related to the theory of radiation.[72]

This rather obscure passage betrays the real nature of Heisenberg's new quantum principle. This was shaped along the lines of the recent Bohr–Kramers–Slater[73] (BKS) project of a quantum theory of radiation, which was meant to be the highest expression of Bohr's correspondence

---

[70] *Ibid.*, p. 293.
[71] The *Ersatzmodell* was nothing but Landé's atomic core vector model that he used for the analysis of the anomalous Zeeman effect, as described in Section 2.1.3.
[72] Heisenberg (1924), p. 299.   [73] Bohr, Kramers, and Slater (1924).

principle. According to the BKS theory, the frequency of the radiation emitted during the transition between stationary states was equal to the frequency emitted by a set of virtual harmonic oscillators. The correspondence was only 'virtual', in the sense that the model was 'a purely logical tool, a theoretical fiction which, though constructed within a conceptual framework irreducible to the quantum theoretical, can nevertheless enable us to explore certain aspects of the reality of the atoms'.[74] Similarly, the correspondence between the classical and the quantum Hamiltonian in Heisenberg's new quantum principle was only a theoretical fiction to explore the anomalous Zeeman effect. As Heisenberg teasingly announced to Pauli, 'despite you, I am going now to publish it (without physical interpretation) with the papal blessing'.[75]

The main tribute to Bohr, the one which led him to concede the 'papal blessing', concerned the building-up principle. Whereas Heisenberg's first core model and the Heisenberg–Landé branching rule violated it, the new quantum principle 'was in a natural connection with the building-up principle ... The building-up principle is directly connected to the ascription of two quantum numbers to the coupling energy'.[76] The riddle of statistical weights was avoided: thanks to a *zweideutig j*, no anomalous decrease in the number of states occurred during the building-up process. The new quantum principle could entail the doublet fine structure and, more generally, the multiplet fine structure, since it entailed the Landé $g$ factors. In spite of that, as Heisenberg readily admitted, the new principle did not explain the anomalous Zeeman effect. It was more a contribution to the mathematics than to the physics of the Zeeman effect. Furthermore, the principle was successfully extended to many-electron systems only in the case of strong magnetic fields, not in the case of weak fields. There were other difficulties and, as Heisenberg knew, 'a theory can still be false when it gives something right, but it can never be right when it gives something false'.[77] The origin of half-integral $k$ for optical doublets was still unclear, especially if compared to the corresponding integral values for relativistic doublets. No unification for optical and relativistic doublets was yet possible. Also the origin of the alleged core magnetic anomaly remained mysterious. Heisenberg simply noticed that the addition of the valence electron with $k = 1/2$ did not seem to affect the magnetic factor 2 of the

---

[74] Petruccioli (1988), English translation (1993), p. 120. See the same chapter for a detailed analysis of the BKS theory.
[75] Heisenberg to Pauli, 8 June 1924. In Pauli (1979), p. 155.
[76] Heisenberg (1924), pp. 301, 303.
[77] Heisenberg to Pauli, 21 February 1923. In Pauli (1979), p. 82.

## 2.2 Bohr, Heisenberg, and Pauli on spectroscopic anomalies

core.[78] But he did not take the further crucial step of associating the magnetic anomaly with the electron's half-integral values. This step was taken by Pauli, who dispensed with the atomic core and finally identified the valence electron as the only culprit of all these anomalies.

### 2.2.4 Pauli: from the electron's *Zweideutigkeit* to the exclusion rule

> Heisenberg's viewpoint sheds no light on the half-integral quantum numbers and on the failure of Larmor's theorem. But I regard such an explanation the most important step and since [Heisenberg's] entire story is purely formal and contains no new physical idea, it is not the theory I hope for.[79]

With these words Pauli dismissed Heisenberg's second core model. Explaining the failure of Larmor's theorem and half-integral values in a fashion that did not land quantum theory in incoherence became Pauli's main goal in the year 1923–4. Following this path Pauli came to anticipate the concept of electron spin and to introduce the 'exclusion rule'.

Pauli spent the winter term 1922–3 in Copenhagen, working with Bohr on a paper about the anomalous Zeeman effect in which no half-integral quantum numbers were used. As Bohr intended, the paper should have offered an analysis of the Zeeman effect in perfect agreement with the orthodox quantum conditions. After many difficulties, Bohr and Pauli withdrew the paper. Yet an echo of Bohr's strenuous attempt to retain integral quantum numbers remained in Pauli's aversion to Heisenberg's models.

In April 1923, Pauli made his first important contribution to the anomalous Zeeman effect.[80] He decided to follow a different route to the problem, one that did not pass through any modelling of empirical data. The upshot was to recover the term values for the anomalous Zeeman effect from the known term values for the Paschen–Back effect via a new rule, since the Paschen–Back effect was more convenient to analyse than the anomalous Zeeman effect. Pauli calculated the $mg$ splitting factors for doublets, triplets, quadruplets, and quintuplets in strong fields where the magnetic quantum number $m$ was set equal to

$$m = m_1 + \mu$$

with $m_1$ taking $(2k-1)$ integral values $0, \pm 1, \ldots, \pm(k-1)$ and $\mu$ taking $2r$ values $\pm 1/2, \ldots, \pm(r-1/2)$ for even multiplets (e.g. doublets), and

---

[78] Heisenberg (1924), p. 300.
[79] Pauli to Kramers, 19 December 1923. In Pauli (1979), p. 135.  [80] Pauli (1923).

$0, \pm 1, \ldots, \pm(r - 1/2)$ for odd multiplets (e.g. triplets). Following the programmatic intent, no model-based interpretation was given for these quantum numbers. Only in a later article (submitted in October 1923)[81] did Pauli explicitly fall back on the atomic core model in interpreting them as

$$m_1 \to m_k \quad \text{and} \quad \mu \to m_r$$

where the first is the orbital magnetic moment and the second the core magnetic moment, so that the magnetic quantum number $m = m_1 + \mu$ denoted the total magnetic moment of the atom. Without explicitly mentioning the atomic core model, in the first article Pauli went on to calculate the Zeeman energy $W = W_0 + mgh\nu_L$ (2.11) for strong magnetic fields (where crucially $g = 1$) as

$$W/h\nu_L = m_1 + 2\mu = m + \mu = 2m - m_1 \qquad (2.22)$$

where once again the puzzling value 2 appeared in the Zeeman energy. The strong-field term values for both even and odd multiplets followed. Most interestingly, once these had been calculated, it was also possible to get the term values for weak fields thanks to a new rule giving the 'symmetry condition for the values of the terms in the transition from strong to weak fields',[82] the so-called 'permanence of the $g$ sums' as Landé dubbed it:[83]

> The sum of the energy values in all those stationary states belonging to given values of $m$ and $k$, remains a linear function of the field strength during an entire transition from weak to strong field.[84]

As Pauli acknowledged, this rule had already been introduced in Heisenberg's first core model, although the particular interpretation that Heisenberg gave to it in terms of the statistical conservation of the angular

---

[81] Pauli (1924).  [82] Pauli (1923), p. 156.  [83] Landé (1923d).
[84] Pauli (1923), p. 162. Notice that in the later semi-classical vector model for the spin–orbit coupling, this rule came to mean that the sum of the $g$ factors, for levels with the same total angular momentum $J$ or magnetic moment $M$, is the same in all field strengths, independent of the coupling scheme used. In the semi-classical vector model, there are two main coupling schemes for the spin $s$ and the orbital angular momentum $l$ of the valence electron: the $LS$ coupling (i.e. the so-called Russell–Saunders coupling) and the $jj$ coupling. In the $LS$ coupling, given two valence electrons, the two $l$ are coupled together to form the resultant $L$ and so also the two $s$ to form the resultant $S$, where in turn $L$ and $S$ are coupled together to give $J$. In the $jj$ coupling, on the other hand, $l_1$ is coupled with $s_1$ to form $j_1$ and $l_2$ with $s_2$ to form $j_2$, where in turn $j_1$ and $j_2$ form the resultant $J$. The $g$ factors change depending on the $LS$ coupling or $jj$ coupling. Yet, according to Pauli's permanence of the $g$-sum rule, for given $L, S, M$ or for given $j_1, j_2, M$ the sum of the $g$ factors remains the same in the passage from strong to weak fields. See White (1934), pp. 189–91.

## 2.2 Bohr, Heisenberg, and Pauli on spectroscopic anomalies 67

momentum was unacceptable. A few short remarks concluded the article expressing Pauli's dissatisfaction with his own result: since the half-integral quantum numbers had not yet been explained, this contribution was purely formal and without any new physical idea, precisely as with Heisenberg. Pauli's main worry remained the violation of Larmor's theorem, as evident from formula (2.22). There was a sort of doubling of the magnetic moment of the atom ($2m$) in the presence of an external field that Pauli, like Landé and Heisenberg, traced back to an alleged core magnetic anomaly ($2\mu$).

In the autumn–winter 1923–4, while Heisenberg was working on his new quantum principle, Pauli was struggling with half-integral quantum numbers and with Larmor's theorem, as he wrote in a letter to Bohr, which is worth quoting in some detail:

> The atomic physicists in Germany fall today into two groups. One group first works through a given problem with half-integral values of the quantum numbers and if it does not agree with experience, then they work with integral quantum numbers. The others calculate first with integral numbers and if it does not agree, they calculate with halves. Both groups however have the characteristic in common that there is no a priori argument to be had from their theories that tells which quantum numbers and which atoms should be calculated with half-integral values ... and which with integral values. They can decide this only a posteriori by comparison with experience. I myself have no taste for this kind of theoretical physics ... I am far more radical than the 'half-integral-number' atomic physicists. This is because I do not believe that the deviation of reality from the results obtained from the theory of periodic systems in many-electron atoms can be explained by just plugging in half-integral numbers in the final formulae of that theory ... But I believe that considerations of this kind can be fruitful only if we manage to put them in direct association with the failure of Larmor's theorem. (The interpretation of this failure through the simple assumption that the ratio between the core magnetic moment and its angular momentum is twice as big as the classical value is too formal. It must be replaced by another interpretation. Unfortunately, with respect to this main point Heisenberg's considerations do not lead us beyond what we already know).[85]

And the search for an interpretation of the failure of Larmor's theorem put Pauli on the right track that neither Bohr nor Heisenberg had envisaged. Pauli first tried to explain the failure of Larmor's theorem by hypothesizing a relativistic correction to the mass of the electrons in the core. In an important letter to Landé,[86] Pauli raised the question as to whether a deviation from the

---

[85] Pauli to Bohr, 21 February 1924. In Pauli (1979), pp. 147–8.
[86] Pauli to Landé, 10 November 1924. In Pauli (1979).

Larmor frequency $\nu_L$ would be expected in classical circumstances, once the effect of the relativistic change of the mass of the electron

$$m = \frac{m_0}{\sqrt{1 - v^2/c^2}} \qquad (2.23)$$

was taken into account. The magneto-mechanical ratio of the atom would accordingly deviate from its normal value

$$\frac{\mu}{p} = \frac{e}{2m_0 c} \qquad (2.24)$$

by a correction factor

$$\gamma = \frac{m_0}{m} = \sqrt{1 - v^2/c^2} \qquad (2.25)$$

that would affect both the Zeeman energy

$$W = \gamma m h \nu_L \qquad (2.26)$$

and the Larmor frequency

$$\nu = \gamma \nu_L = \nu_L \left\{ 1 + \frac{\alpha^2 Z^2}{\left[n - k + \sqrt{k^2 - \alpha^2 Z^2}\right]^2} \right\}^{-1/2} \qquad (2.27)$$

where $\alpha$ is the fine structure constant $\alpha = 2\pi e^2/hc$ and $Z$ is the atomic number. When $n = k = 1$ (i.e. in the case of the hydrogen atom), $\gamma$ would reduce to

$$\gamma = \sqrt{1 - \alpha^2 Z^2} \qquad (2.28)$$

with $\alpha^2 Z^2$ so small (since $Z = 1$) that the correction factor would not differ significantly from 1, and hence no influence on the Zeeman splitting would be observed. But for high $Z$, $\gamma$ would supposedly differ considerably from 1, and it would be possible to detect the violation of the Larmor frequency. Pauli then went on to calculate the eventual influence of $\gamma$ on the Zeeman effect for elements with high $Z$. Although he had the strong feeling that 'the calculated deviations of the Zeeman effect from the usual ones are not actually available',[87] he asked Landé to let him know whether any relativistic correction could actually be detected. Anticipating an eventual negative answer, Pauli concluded:

---

[87] Ibid., p. 171.

## 2.2 Bohr, Heisenberg, and Pauli on spectroscopic anomalies

> What would then follow in the case that the calculated deviations were not effectively detected (as I think it is likely to be)? The entire calculation relies on the assumption that the $K$-shell [i.e. the core] ... possesses a non-vanishing resultant angular momentum. But this momentum of the $K$-shell is not to be taken too seriously and has something unrealistic about it ... (I am tempted to say that not only has the core momentum half-integral values, but also a half-integral reality) ... A future, adequate (not *à la* Heisenberg) theory of the anomalous Zeeman effect must give reasons for the non-availability of the relativistic correction factor.[88]

In the following few days Landé, from the Institute for Physics at Tübingen, informed Pauli that the observed relativistic correction factor in cadmium amounted to 6% and in zinc to 2%. But Pauli was

> not quite satisfied with these results ... One must consider elements with higher atomic numbers ... in order to decide the question with certainty ... It is a pity that the Zeeman effect is not precisely measured in alkalis with higher atomic numbers. Do we know the precise values of the Ba spark spectrum? And those in the III column? I am convinced since the beginning that there is no relativistic effect.[89]

Landé finally announced the negative result for thallium ($Z = 90$) that Pauli welcomed as 'totally sufficient to eliminate the last part of a possible doubt about the empirical state of affairs'.[90] The absence of a relativistic correction in thallium spoke against the assumption of an atomic core allegedly endowed with a non-zero angular momentum. Against this assumption stood other facts that Pauli listed in the same important letter to Landé on 24 November 1924. For instance, in the observable characteristics, the K-shell did not differ from the closed shells with higher quantum numbers. But the K-shell was typically ascribed a non-vanishing angular momentum, whereas the closed shells with higher quantum numbers were ascribed a zero angular momentum. This introduced an asymmetry in the theory, to which nothing corresponded at the empirical level. Pauli drew the following remarkable conclusion:

> In the alkalis, the valence electron alone makes the complex structure as well as the anomalous Zeeman effect. The contribution of the atomic core is out of question (also in other elements). In a puzzling, non-mechanical way, the valence electron manages to run about in two states with the same $k$ but with different angular momenta.[91]

---

[88] *Ibid.*, p. 171.   [89] Pauli to Landé, 14 November 1924. In Pauli (1979), p. 172.
[90] Pauli to Landé, 24 November 1924. In Pauli (1979), p. 176.   [91] *Ibid.*, p. 177.

The puzzling, non-mechanical way in which the valence electron manages to run about in two different states with the same orbital angular momentum $k$ was referred to as the electron's *Zweideutigkeit* in an article submitted one week later (on 2 December 1924), in which Pauli summarized his unsuccessful search for the relativistic correction:

> From this viewpoint the doublet structure of the alkali spectra as well as the failure of Larmor's theorem arise through a specific, classically non-describable sort of 'twofoldness' [*Zweideutigkeit*] of the quantum-theoretical properties of the valence electron.[92]

The concept of *Zweideutigkeit* – as we have seen in Section 2.2.3 – originally sat squarely within the atomic core model as the 'twofoldness', i.e. the 'two-quantum-numbered' (*Zahlenpaar*) angular momentum of the core, electron, and atom. As such, it underpinned Heisenberg and Landé's branching rule. One year later, Pauli reinterpreted this very same concept as the non-mechanical ability of the valence electron to run about in two different states with the same orbital angular momentum. As the following scientific developments revealed, this was nothing but the two-valuedness of the spin angular momentum. Pauli did not speak in terms of spin in 1924, but he was the first to abandon the time-honoured atomic core model in the explanation of complex spectra and of the anomalous Zeeman effect. He was the first to regard the valence electron as alone responsible for these spectroscopic anomalies in the light of its enigmatic, non-mechanical *Zweideutigkeit*.

In the same letter to Landé, Pauli ascribed three quantum numbers to the valence electron along the lines of the Bohr–Coster theory of X-rays:[93] $n_{k_1,k_2}$ where $n$ was the principal quantum number, $k_1$ the azimuthal quantum number and $k_2$ denoted the magnitude of the relativistic correction. Pauli went on to introduce another quantum number, the magnetic number $m$, under the assumption that the atom was placed in a strong magnetic field. Given the null experimental result about the relativistic correction, Pauli then replaced $k_2$ with the magnetic quantum number $m_2$, and $m$ with $m_1$, where the number $m_2$ represented the total magnetic interaction energy with the strong field, with '$m_1$ half-integral; why, not yet clear; $m_2 = m_1 \pm 1/2$'. In so doing, Pauli could

---

[92] Pauli (1925a), p. 385.
[93] The Bohr–Coster theory of X-rays (1923) improved Sommerfeld's earlier relativistic treatment by identifying any X-ray term with three quantum numbers $n_{k_1,k_2}$, with $\Delta k_1 = \pm 1$, and $\Delta k_2 = \pm 1,0$. Landé (1923b) identified $k_2$ with the inner quantum number $j$ of the alkali doublets, because it could take two states $k_2 = k_1, k_1 - 1$ as $j$ did. In so doing he took an important step towards the unification of relativistic and optical doublets.

catch a glimpse of a logical connection between the number 2 of these states and the half-integral quantum numbers, on the one side, and the violation of Larmor's theorem on the other side.[94]

He imagined that a strong magnetic field could be associated with any atom so that the Paschen–Back states were realized, and every electron could accordingly be characterized by the four quantum numbers $n_{k_1,m_1,m_2}$. The first consequence of this assumption was that 'Bohr's *Zwang* ... disappears'.[95] No atomic core angular momentum, no *unmechanischer Zwang* in the alleged core–electron coupling. It was then the turn of the Landé–Heisenberg branching rule:

> From a free atomic core with $N_i$ possible states springs a term-system with $(N_i + 1)(2k_1 - 1)$ states and one with $(N_i - 1)(2k_1 - 1)$ in the presence of a field... Now I interpret these total $2N_i(2k_1 - 1)$ states simply as follows: the $N_i$ states of the atomic core remain, and the valence electron takes $2(2k - 1)$ states as in the alkalis.[96]

No core angular momentum, no branching states with the addition of an outer electron: the doubling of the number of states was attributed to the electron's *Zweideutigkeit*. The riddle of statistical weights could then be averted: 'The *Aufbauprinzip* is strictly valid in my case. This fact seems to me to make the here suggested viewpoint so superior over those held so far, that despite all difficulties I regard it as the physically more correct'.[97]

But even more interesting for our story was what Pauli announced as 'the second strength of my viewpoint', namely the possibility of classifying equivalent orbits 'in the most natural way'.[98] He referred back to a recent work of Edmund Stoner, a Ph.D. student of Rutherford at Cambridge, who a few months earlier had published an article in the *Philosophical Magazine*[99] on the electronic distribution in atomic shells. Stoner's insight was that the largest possible number $N$ of electrons in a closed $n_{kj}$ shell coincided with the number $N$ of sublevels into which the term $n_{kj}$ split in a weak magnetic field. Using this rule, Stoner could account for the electronic distribution in the periodic table (2, 8, 18, 32, ...): the $1_1$ group was closed with helium, the $2_{11}$ group was closed with beryllium, the $2_{21}$ with calcium, and the $2_{22}$ with neon. Not only was Stoner's rule better than Bohr's rival building-up schema, as it gave a more natural classification of

---

[94] Pauli to Landé, 24 November 1924. In Pauli (1979), p. 178.   [95] *Ibid.*, p. 178.
[96] *Ibid.*, p. 178.   [97] *Ibid.*, p. 179.
[98] *Ibid.*, p. 180.
[99] Stoner (1924). On Stoner's work and its relevance to the origin of the exclusion principle see the excellent article of Heilbron (1982), pp. 280–7.

subshells without arbitrary reopenings. But it also accounted successfully for some evidence about the spectrum of ionized carbon that patently violated Bohr's schema.[100]

Pauli reinterpreted Stoner's rule in the light of the electron's *Zweideutigkeit*. Having dispensed with the atomic core, Pauli could abandon the inner quantum number *j* that Stoner had used. In the presence of a strong magnetic field, the number of sublevels into which the spectral term split was known to be $2(2k-1)$, i.e. $2n^2$. But $2n^2$ was also Rydberg's 'cabalistic'[101] formula – as Sommerfeld called it – for the *n* periods of Mendeleev's table, in which the number of electrons per period (2, 8, 18, ...) were intended as $2 \times (n=1)^2$, $2 \times (n=2)^2$, $2 \times (n=3)^2$. Hence, surprisingly enough, the number of states an electron can take up in a strong magnetic field coincided with the number of states an electron can take up in the *n*-th period of the periodic table. This was justified by Pauli's permanence of *g* sums: the sum of the energy-states associated with a given magnetic quantum number *m* remains the same in all field strengths. So it remains the same in the transition from strong fields (Paschen–Back effect), to weak fields, up to zero fields (normal atoms, following Rydberg's rule for electronic distribution). Having so reinterpreted Stoner's rule, Pauli traced back the lengths of the periods to what he presented as a prescriptive rule [*Vorschrift*]:

> I can trace back the closure of groups [i.e. in Stoner's schema] ... to a single prescription that seems to me extremely natural. I am thinking of a so strong magnetic field that all electrons can be characterised through the symbol $n_{k_1,m_1,m_2}$ as above described. Then it should be forbidden that more than one electron with the same (equivalent) *n* belongs to the same values of the three quantum numbers $k_1$, $m_1$, $m_2$. When an electron corresponds to a given $n_{k_1,m_1,m_2}$-state, this state is occupied.[102]

There was still a long way to go for this prescriptive rule to become the exclusion principle. As we shall see in Chapter 4, the process that led to promoting this rule to the rank of a scientific principle is linked to the development of quantum mechanics as a theoretical framework that

---

[100] 'The strongest evidence for Stoner came from Alfred Fowler's finding that the spectrum of ionized carbon shows doublets that, because of their appearance and position, had to arise from a $2_2$ term. The natural interpretation of this unexpected oddity was, as Bohr himself explained it, that "the singly charged carbon ion in its normal state besides two electrons in $2_1$ orbits possesses one electron in a $2_2$ orbit". That interpretation required a correction in the interpretation he had laid down: the fifth electron for some reason does not take up the $2_1$ path he [Bohr] had prepared for it. The reason, according to Stoner: since only two $2_1$ orbits exist, the fifth electron necessarily falls into a $2_2$ circle.' Heilbron (1982), pp. 283–4.

[101] On Rydberg's formula, see Pauli (1955).

[102] Pauli to Landé, 24 November 1924. In Pauli (1979), p. 180.

Pauli's rule came to be built into from the ground up. Hence, in this original historical context, it is more appropriate to refer to it as a rule than as a principle. Pauli himself called it *Ausschließungsregel*[103] or *meine Ausschlußregel* (exclusion rule),[104] while Heisenberg teasingly called it Pauli's *Verbot der äquivalenten Bahnen* (Pauli's prohibition of equivalent orbits).[105]

Pauli admitted that 'we cannot give a closer foundation to this rule, yet it seems to present itself in a very natural way'.[106] The exclusion rule was then introduced as a theoretical consequence of reinterpreting Stoner's rule in the light of the electron's *Zweideutigkeit*, with the help of the permanence of the *g*-sum rule. The closure of electronic groups in the periodic table followed naturally from it. Furthermore, the rule also accounted for the possible combinations of electrons in non-closed shells. For instance, it finally became clear why – despite alkaline earths containing two s-electrons in their ground state, which could give rise to singlets and triplets – only singlets were usually observed. The reason is that the two s-electrons are equivalent and hence certain terms, namely triplets, are forbidden according to Pauli's rule. Only when one of the two electrons is excited to an s-orbit of different $n$, does the triplet appear: similarly, for any two equivalent p-electrons or d-electrons. Out of all logically possible combinations of any two electrons, only a few of them can actually be realized: those in which no two equivalent electrons are present. Everything seemed to fit Pauli's rule very well. Only Bohr's correspondence principle was left out: how to reconcile the classical periodic motions presupposed by the correspondence principle with the classically non-describable *Zweideutigkeit* of the electron's angular momentum? Pauli could not help remarking that 'this is indeed a very strange state of affairs'.[107]

## 2.3 The turning point

Bohr welcomed Pauli's exclusion rule, although he did not hide his perplexity about the classically non-describable nature of the *Zweideutigkeit* and the impossibility of reconciling it with the correspondence principle:

---

[103] Pauli to Landé, 15 December 1924. In Pauli (1979), p. 191.
[104] Pauli to Landé, 25 December 1924. In Pauli (1979), p. 196.
[105] Heisenberg to Pauli, 16 November 1925. In Pauli (1979), p. 256.
[106] Pauli (1925b), p. 776.   [107] Pauli to Landé, 24 November 1924. In Pauli (1979), p. 180.

74                    *2 The origins of the exclusion principle*

> Dear Pauli,
> I cannot easily describe how welcome your submission was. We are all excited for the many new beautiful things you have brought to light. I do not need to advance any criticism, since you by yourself, better than anyone else could have done, have characterised the whole thing in your letter as complete madness... I am not very sure either whether you are overstepping a dangerous threshold – intoning your old '*Carthaginem esse delendam*' – when you declare the definitive death sentence about a correspondence-like explanation of the closure of groups... I sense that we are here standing at a decisive turning point.[108]

The *Zweideutigkeit* stood indeed at a decisive turning point, as Bohr noticed. It introduced a new crucial property for the electron, which escaped any classical description as much as the exclusion rule eluded any quantum theoretical proof. The theoretical breakthrough called for experimental confirmation. On 9 January 1925 Pauli visited Landé at the Institute for Physics in Tübingen, which was the leading centre for spectroscopic research. The data on the spectrum of lead ($Z = 82$) turned out to be in striking agreement with Pauli's rule.[109] Intense discussions followed within the small *Gesellschaft* of physicists gathered at Landé's house till late at night. Among them was a young Ph.D. graduate from Columbia University, Ralph Kronig, who had the chance to read Pauli's letter to Landé where the new rule was announced. As Kronig recalled many years later:

> Pauli's letter made a great impression on me and naturally my curiosity was aroused as to the meaning of the fact that *each individual* electron of the atom was to be described in terms of quantum numbers familiar from the spectra of the alkali atoms, in particular the two angular momenta $l$ and $s = 1/2$ encountered there. Evidently $s$ could now no longer be attributed to a core, and it occurred to me immediately that it might be considered as an intrinsic angular momentum of the atom. In the language of the models which before the advent of quantum mechanics were the only basis for discussion one had, this could only be pictured as due to a rotation of the electron about its axis ... the same

---

[108] Bohr to Pauli, 22 December 1924. In Pauli (1979), pp. 194–5.
[109] 'Lead in its normal state has two superficial p-electrons (electrons with $k = 2$). Five possibilities therefore exist for the lowest state ... There result five $n_{kj}$ terms defined by values of $j$ of 2, 2, 1, 0, 0, a prescription testable by calculating the $g$ values and examining the anomalous Zeeman effect in lead. Pauli was not pleased to find that only four levels had been identified, and that they had been assigned $j$'s of 1, 2, 1, 0. If the experimentalists had not erred, he wrote Landé, "my closure rule must be modified for complicated cases" (Pauli to Landé, 15 and 25 December 1924). Heroic measures were called for. Pauli stopped at the dull town of Tubingen en route to Hamburg from Vienna, where he had spent Christmas. There, on Landé's kitchen table, he examined Back's latest photographs of the lead-spectrum in a magnetic field. Analysis disclosed five terms with the predicted $j$ values.' Heilbron (1982), p. 305.

afternoon, still quite under the influence of the letter I had read, I succeeded in deriving with it the so-called relativistic doublet formula.[110]

Kronig introduced the idea of electron spin. If a spinning or self-rotating electron is associated with a magnetic moment of one Bohr magneton, via a Lorentz transformation, in the rest frame of the electron the electric field created by the protons of the nucleus would appear as a magnetic field, with which the electron's magnetic moment would interact.

Since the idea of a spinning electron was still based on a semi-classical mechanical model, given Pauli's reluctance to use semi-classical models, he dismissed it with the remark 'this is indeed quite a witty idea'.[111] Kronig, disheartened, abandoned the idea himself. In the meantime, on 16 January, Pauli submitted the paper in which the exclusion rule was announced. The article ended with an apologetic remark on the impossibility of reconciling the new view with the correspondence principle. As an intellectual tribute to Bohr, Pauli advanced the hope that 'in the near future, a fusion of these two viewpoints [i.e. *Zweideutigkeit* and correspondence principle] will be achieved'.[112] Yet Pauli remained deeply convinced that no classical description for the electron's *Zweideutigkeit* could ever be given. Indeed, the subsequent developments did justice to his intuition.

The 'witty idea' was resurrected nine months later by two Dutch physicists, George Uhlenbeck and Samuel Goudsmit, who – independently of Kronig – arrived at a similar conclusion starting from Pauli's work. Uhlenbeck and Goudsmit published a short note in *Naturwissenschaften* where the fourth degree of freedom for the electron was identified with a rotation of the electron about its axis.[113] The gyromagnetic ratio for this new degree of freedom was twice the corresponding ratio for the orbital motion, and the coupling between the orbital angular momentum vector and the intrinsic angular momentum vector was invoked as the key for an understanding of doublets.

The main obstacle the spinning electron model faced was a discrepancy between the observed doublet splitting and the calculated fine structure: a factor of 2 had already appeared in Kronig's calculation. Note that this factor had nothing to do with the well-known value 2 of the Landé *g* factor: an explanation of the latter could be given in Uhlenbeck and Goudsmit's model, where the gyromagnetic ratio for a self-rotating electron with surface charge turned out to be equal to $2 \cdot e/(2mc)$. But an explanation of this other factor 2 appearing in Kronig's calculation proved more difficult.

[110] Kronig (1960), pp. 19–20.  [111] *Ibid.*, p. 21.  [112] Pauli (1925b), p. 771.
[113] Uhlenbeck and Goudsmit (1925).

Heisenberg repeated Kronig's calculation in November 1925, and in a letter to Pauli he himself expressed doubts about the model.[114] Despite this and other difficulties,[115] Bohr adhered to the spinning electron model because it restored a desirable, classically describable picture. Bohr's enthusiasm overcame Heisenberg's reluctance,[116] but not Pauli's, who firmly rejected 'the new heresy'.

In February 1926 a solution was finally found for the puzzling factor 2. As Bohr announced to Pauli,[117] a young English physicist, Llewellyn Hillet Thomas, who had spent the previous half year in Copenhagen, had discovered that this factor was simply due to a mistake in the calculation of the relative motion of the electron and the atomic nucleus: an additional angular velocity of the nucleus, due to the effect of special relativity in the rest frame of the electron, should be introduced in the calculation of the equation of motion of the electron's magnetic moment. A copy of Thomas's article[118] was enclosed with the letter.

Pauli did not welcome the new result favourably. In his reply to Bohr[119] he insisted that the question could not be solved the way Thomas suggested and he even asked Bohr to block the publication of Thomas's paper. But Bohr remained firm in his position: 'Dear Pauli, ... your letter has only strengthened our belief in the validity and justification of the argument.'[120] Pauli was left with no other choice than 'capitulate completely'.[121] Pauli's capitulation coincided with the publication of a second article by Uhlenbeck and Goudsmit,[122] where the spinning electron model was deployed to calculate the hydrogen spectrum, while in May a confirmation of the validity of Thomas's calculation came from the work of Frenkel.[123]

As later developments revealed, Pauli's reluctance towards the spinning electron model contained a kernel of truth. The classically non-describable

---

[114] Heisenberg to Pauli, 24 November 1925. In Pauli (1979), p. 265.
[115] As Uhlenbeck later recalled, 'It was quite clear that the picture of the rotating electron, if taken seriously, would give rise to serious difficulties. For one thing, the magnetic energy would be so large that by the equivalence of mass and energy the electron would have a larger mass than the proton, or, if one sticks to the known mass, the electron would be bigger than the whole atom! In any case, it seemed to be nonsense.' Quotation from van der Waerden (1960), p. 214.
[116] 'Bohr's optimism about Goudsmit's theory has so much influenced me, that I'd really like to believe in the magnetic electron.' Heisenberg to Pauli, 24 December 1925. In Pauli (1979), p. 271.
[117] Bohr to Pauli, 20 February 1926. In Pauli (1979), p. 295.   [118] Thomas (1926), (1927).
[119] Pauli to Bohr, 26 February 1926. In Pauli (1979), p. 297.
[120] Bohr to Pauli, 9 March 1926. In Pauli (1979), p. 309.
[121] Pauli to Bohr, 12 March 1926. In Pauli (1979), p. 310.
[122] Uhlenbeck and Goudsmit (1926).   [123] Frenkel (1926).

## 2.3 The turning point

electron's *Zweideutigkeit* could not be cast in classical terms, and the electron's spin turned out to have no classical analogue. In 1928 Dirac finally showed that

> the incompleteness of the previous theories [i.e. Pauli's and Uhlenbeck–Goudsmit's] [lay] in their disagreement with relativity... All the same there is a great deal of truth in the spinning electron model, at least as a first approximation.[124]

The Dirac relativistic wave equation for the electron finally allowed the derivation of the spin magnetic moment, and in so doing vindicated Pauli's original reluctance to cast the electron's *Zweideutigkeit* in the semi-classical spinning electron model. But that is another story, and I shall come back to it in Chapter 4.

---

[124] Dirac (1928a), p. 610.

# 3

# From the old quantum theory to the new quantum theory: reconsidering Kuhn's incommensurability

The transition from the atomic core model to the electron's spin, the debate on the adequacy of semi-classical models, and the lack of an appropriate scientific terminology are symptomatic of the revolutionary transition from the old quantum theory to the new quantum theory around 1921–5. As such, they provide us with a foil for rethinking Kuhn's view on scientific revolutions. In this view, scientific revolutions are distinctively accompanied by incommensurability between paradigms, or – to use Kuhn's later terminology – untranslatability between scientific lexicons. In the light of the historical reconstruction offered in Chapter 2, I shall here argue for the prospective intelligibility of the revolutionary transition around 1924 via a two-step argument that (1) reconsiders Kuhn's notion of incommensurability as untranslatability (Section 3.2), and (2) offers a positive account of the way the electron's *Zweideutigkeit* and the exclusion rule came out of the old quantum theory (Section 3.3).

## 3.1 The revolutionary transition from the old quantum theory to the new quantum theory

On 5 November 1980, Thomas Kuhn delivered a lecture at Harvard University entitled 'The crisis of the old quantum theory: 1922–25'.[1] In his distinctive style of reasoning, Kuhn presented the rise of quantum mechanics after 1925 as the result of a period of crisis of the old quantum theory between 1922 and 1925. The old quantum theory – in Kuhn's

---

[1] The HUSC Research Lecture was video-recorded, and it is now held at the Godfrey Lowell Cabot Science Library of the Harvard College Library. I thank the staff of the Cabot Science Library for the inter-library loan of this videotape.

view – cannot be regarded as a full-blown theory but rather as a set of algorithms to solve problems and paradoxes. It was a mixture of rules of thumb, numerology, and lower-level phenomenological laws without a theoretical foundation. On the other side, if we consider that for almost a decade physicists learnt to live with paradoxes and developed sophisticated techniques to solve an increasing number of problems, the old quantum theory can be regarded as a very fine theory. At some point, between 1922 and 1924, this theory proved unable to deal with an increasing number of experimental anomalies, namely the X-ray doublets, the alkali doublets, and the anomalous Zeeman effect, among many others that I did not discuss in Chapter 2, such as the helium atom and dispersion theory. The old quantum theory – Kuhn continues – cried out for fundamental changes that via Landé's work on the anomalous Zeeman effect, and Kramers's research on dispersion theory, among many other important steps, led finally to the new quantum theory.

Pauli's exclusion rule, which Kuhn did not discuss in his lecture, was the genuine product of this period of crisis and revolutionary transition from the old (pre-1925) quantum theory to the new (post-1925) quantum theory. Trapped as it was between the waning fortunes of the old quantum theory, and the not-yet-developed new quantum theory, the exclusion rule was one among many phenomenological laws available at the time. Yet, by contrast with other phenomenological laws of the old quantum theory, which were used as a ladder to be thrown away once one had climbed up it, the exclusion rule remained in the new quantum theory. Indeed, it was built into the new quantum theory from the ground up, as we shall see in Chapter 4. Before turning to analyse the specific role that Pauli's rule played in the new quantum theory, in this chapter I want to take a closer look at the revolutionary transition around 1922–5 that led to the introduction of the electron's spin and of Pauli's rule.

I shall argue that Pauli's rule is the product of the revolutionary transition from the atomic core model to the electron spin model, without however being itself the element that warrants the rational continuity of this revolutionary transition. This suggests a picture of rational continuity that differs from Michael Friedman's. As discussed in Section 1.3, Friedman traces the rational continuity of revolutionary transitions back to the specific bridging role of relativized a priori principles. His analysis of Einstein's light principle and equivalence principle suggests that both originated from well-established empirical facts that at some point were 'elevated' to the status of constitutive a priori principles (in Reichenbach's sense) for the new relativity framework. But Pauli's principle does not fit this picture. There was no well-established empirical fact at its origin. Nor, as will become clear in Chapter 4, was this phenomenological law 'elevated' to the status of a constitutive principle for the new quantum theory framework.

Rather, at the origin of Pauli's rule there was a series of well-known spectroscopic anomalies that could not be reconciled with the theoretical assumptions of the old quantum theory. Echoing Kuhn, Pauli's rule was one of the products of the crisis of the old quantum theory, of its inability to make sense of an increasing number of anomalies. We cannot appeal to the exclusion rule to warrant the rational continuity between the old and the new quantum theory, because Pauli's rule itself (with the related concept of the electron's *Zweideutigkeit*) was a brand new product of this scientific revolution.

However, rational continuity can still be maintained, albeit at a different level. The revolutionary transition around 1922–5 did not amount to Kuhnian incommensurability, as I shall argue in Section 3.2. The new framework or, to use Kuhn's later terminology, the new scientific lexicon that emerged after 1924 included Pauli's exclusion rule and the electron spin. It emerged by dismantling some key concepts of the old lexicon, such as the atomic core with a non-zero angular momentum; the strict validity of Larmor's theorem; and, more generally, the tendency to shape quantum theory as closely as possible along the lines of classical physics. The clash between Pauli's rule and Bohr's correspondence principle was not a temporary setback. As Pauli was well aware by 1924, this clash pointed instead to a deeper conflict between classical physics and quantum theory. It pointed to the impossibility of describing quantum properties in classical terms. The electron's *Zweideutigkeit* did not lend itself to classical interpretation, as the controversy on the spinning electron model revealed. And, despite Bohr's enthusiasm for the apparent reconciliation achieved by Uhlenbeck and Goudsmit's model, the very near future gave the verdict to Pauli's original reluctance to cast the electron's *Zweideutigkeit* in semi-classical terms. The electron spin turned out to have no classical analogue and it could be justified only within the new quantum theory, with Dirac's relativistic wave equation for the electron in 1928, as we shall see in Chapter 4.

Despite the revolutionary context, as the lack of an adequate scientific terminology for the new concepts testifies, no threat of incommensurability looms here on the horizon. On the contrary, there is common ground, a territory where the old quantum theory partially overlapped with the new quantum theory. This is the same territory whence Pauli's rule originated. As I shall clarify in Section 3.3, it is the interplay between theoretical assumptions of the old quantum theory *and* anomalous phenomena that jointly led to the electron's *Zweideutigkeit* and to Pauli's rule and that in so doing warranted the rational continuity and prospective intelligibility of this revolutionary transition. The electron's *Zweideutigkeit* and the exclusion rule were the results of a gradual stretching process of the old quantum theory that, in order to achieve a better fit with anomalous phenomena, made the theory finally unrecognizable.

## 3.2 Reconsidering Kuhnian incommensurability

### 3.2.1 Kuhn on scientific lexicons: incommensurability as untranslatability

It is now well over forty years since Thomas Kuhn put forward his groundbreaking view of scientific revolutions as the result of a period of crisis that any well-established scientific paradigm must face after a possibly long period of so-called 'normal science'. Kuhn famously characterized scientific revolutions in terms of 'incommensurability' between paradigms.[2] The notion of incommensurability sits squarely within a view of science that seeks a shift away from the logical positivist view of scientific progress. Kuhn challenged the view that science is a cumulative acquisition of knowledge, where later theories would constitute a progressive shift with respect to earlier theories. By portraying science as a recursive cycle of normal science, crisis, and revolution, Kuhn offered a radically new picture, according to which later theories in mature science can no longer be taken as more likely to be true than earlier theories. Although after the publication of *The Structure of Scientific Revolutions* Kuhn attempted to mitigate his position, scientific rationality remained on shaky grounds in this view. The five characteristics that Kuhn indicated in *The Essential Tension*[3] as providing the shared basis for theory-choice (accuracy, consistency, broad scope, simplicity, fruitfulness) were considered insufficient to determine why scientists opt for one theory rather than another in any given historical context. They are desirable features of any scientific theory, but they cannot give the verdict to one theory rather than another. Incommensurability between paradigms is the feature primarily responsible for this impasse.

In the original formulation, incommensurability means that any two paradigms are literally not commensurable: they lack a 'common measure' for rational choice, because they do not share the same scientific concepts,

---

[2] Kuhn (1962). [3] Kuhn (1977), pp. 320–39.

methodology, technological resources, and even system of values. Accordingly, inter-paradigm shifts resemble Gestalt switches: practitioners of different scientific paradigms would not only work with different theories, but they would actually 'live in different worlds'. In the absence of a common ground, the challenge consists in explaining how the transition from one paradigm to another is conceivable in the first place. Indeed, if the new paradigm is unintelligible from the viewpoint of the older paradigm – because it resorts to brand new concepts, methodology, values, and so forth – how could practitioners of the older paradigm ever be in the position of embracing the new one?

In his later writings, Kuhn mitigated the drastic picture that the original notion of incommensurability delivered. He reinterpreted incommensurability as untranslatability between scientific lexicons. This reinterpretation was congenial to Kuhn's 'linguistic turn' in the 1980s: applied to the conceptual vocabulary of a scientific theory, the term 'incommensurability' came to mean 'no common measure' intended as 'no common language'. Incommensurability amounted then to the claim that there is no language, neutral or otherwise, into which both theories, conceived as sets of sentences, can be translated without residue or loss.[4] Untranslatability is a much weaker notion than the earlier notion of incommensurability because it is restricted to the language in which theories are formulated and confined to some locally untranslatable terms. Yet the challenge originally posed by incommensurability remains. Untranslatable lexicons still challenge the prospective intelligibility of scientific revolutions. How could speakers of the older lexicon ever acquire a brand new lexicon that is not translatable into their own? It is this challenge that I want to address and possibly reassess in this chapter. But first let us take a closer look at Kuhn's view.

Kuhn defined a scientific lexicon as the conceptual vocabulary of a scientific theory, consisting of 'kind terms' of various sorts: natural kind terms such as 'electron', 'gold', 'water', etc.; artifactual kind terms such as 'microscope', 'nuclear reactor', etc.; social kind terms such as 'nuclear physicist', 'engineer', 'experimenters', etc., among other possible kind terms. All kind terms obey what Kuhn presented as the *no-overlap principle*:

> no two kind terms, no two terms with the kind label, may overlap in their referents unless they are related as species to genus. There are no dogs that are also cats, no gold rings that are also silver rings, and so on: that's what makes dogs, cats, silver, and gold each a kind.[5]

---

[4] Kuhn (1983). Reprinted in Kuhn (2000), p. 36.
[5] Kuhn (1991). Reprinted in Kuhn (2000), p. 92.

Kind terms bring along with them what Kuhn called normic and nomic generalizations. The former are generalizations concerning the kind at issue that admit exceptions; the latter are, on the contrary, exceptionless: 'in the sciences, where they mainly function, these generalizations are usually laws of nature.'[6] Nomic generalizations may in turn be responsible for a stronger form of untranslatability. The same no-overlap principle bars practitioners of a lexicon from importing some of the laws of another lexicon that may be ineffable, or unavailable for conceptual or observational scrutiny within their lexicon.[7] Thus the no-overlap principle acts at two different levels in forbidding any overlap between kind terms belonging to the same contrast set, and in forbidding any overlap between nomic generalizations related to kind terms of different lexicons.

A scientific lexicon has two distinctive features: it is *holistic* and *structured*. The main terms of a lexicon are typically interdefined and are acquired as a whole, together with some relevant laws of nature that provide the conditions of possibility of the lexicon itself. In other words, kind terms are learned together (holistic nature of lexical acquisition) because they are jointly instantiated in situations that exemplify laws of nature. For example, in Newtonian mechanics the main terms 'force', 'mass', and 'acceleration' are interdefined and acquired together with Newton's second law. Kuhn considered Newton's second law as a law-sketch that must be rewritten in different symbolic forms depending on the specific problems it is applied to (e.g. free fall, pendulum, coupled harmonic oscillators).[8] Physics students learn how to apply Newton's second law in different situations by a process like ostension, i.e. by being exposed to a series of exemplary situations, among which they learn to recognize similarity–dissimilarity relationships. These are the relationships that allegedly define the taxonomic structure of the Newtonian lexicon: physics students learn that in the Newtonian lexicon free fall is an example of 'forced' motion (instead of 'force-free' motion as it was for Aristotelians). As such, it is subject to a suitable symbolic expression of Newton's second law. These are the nomic expectations that the term 'force' brings along with it as a projectible term. That is why these terms are not translatable into the language of a physical theory such as Aristotle's or Einstein's, in which Newton's version of the second law does not apply. To learn any one of these three different theories, the interrelated terms must be learned or

---

[6] Kuhn (1993), p. 316.   [7] *Ibid.*, p. 336.
[8] See Kuhn (1970). Reprinted in Kuhn (2000) p. 169.

relearned together, they cannot simply be rendered individually by translation.[9]

Not only are terms acquired together, they are also acquired according to a certain 'lexical structure'. For instance, in Newtonian mechanics, starting from 'inertial mass' and Newton's second law, the introduction of the gravitational law as an empirical regularity smoothes the path to the introduction of the term 'weight' as a relational property that depends on the presence of two or more bodies and on whether they are located, for example, at the surface of the Earth or of the Moon. Vice versa, if we start from 'gravitational mass' and the law of gravitation, 'weight' can be explained as a relational property resulting from gravitational attraction, hence Newton's second law can be empirically introduced as the relation between applied force and the acceleration of a mass measured by gravitational means. In Kuhn's words:

> The two routes thus differ in what must be stipulated about nature in order to learn Newtonian terms, what can be left to instead for empirical discovery. On the first route, the second law enters stipulatively, the law of gravitation empirically. On the second, their epistemic status is reversed. In each case one, but only one, of the laws is, so to speak, built into the lexicon. I do not quite want to call such laws analytic, for experience with nature was essential to their initial formulation. Yet they do have something of the necessity that the label 'analytic' implies. Perhaps 'synthetic a priori' comes closer.[10]

This view of Newton's laws constitutively built into the lexicon sits squarely with Kuhn's declared 'post-Darwinian Kantianism'. Inspired by Friedman's dynamic Kantianism, in his later years Kuhn came to compare his notion of a structured lexicon with relativized a priori principles, intended as principles 'constitutive of *possible experience* of the world ... constitutive of the infinite range of possible experiences that might conceivably occur in the actual world to which they give access'.[11] In his later writings, Kuhn gave an explicitly Kantian twist to his view in the sense that 'like the Kantian categories, the [scientific] lexicon supplies preconditions of possible experience. But lexical categories, unlike their Kantian forebears, can and do change, both with time and with the passage from one community to another.'[12] This is Kuhn's *soi-disant* post-Darwinian Kantianism.[13]

---

[9] See Kuhn (1983).Reprinted in Kuhn (2000), p. 44.   [10] Kuhn (1990), p. 306.
[11] Kuhn (1993), pp. 331–2.   [12] Kuhn (1991), p. 12.   [13] See Lipton (2001).

## 3.2 Reconsidering Kuhnian incommensurability

Thus, as Kuhn repeatedly stressed, acquiring a new scientific lexicon is equivalent to learning a new language: it requires bilingualism, not translatability. Nor does bilingualism imply translatability: once a new lexicon is acquired, scientists will not be able to translate from the newly acquired lexicon to the old one with which they were raised, not even if the new lexicon is expanded with the addition of selected terms from the old lexicon.[14]

There are two main assumptions underpinning the untranslatability thesis. The first is the identification of lexical structures with lexical taxonomies. On this crucial point, it is worth quoting Kuhn in some detail:

> A lexical taxonomy of some sort must be in place before description of the world can begin. Shared taxonomic categories, at least in an area under discussion, are prerequisite to unproblematic communication, including the communication required for the evaluation of truth claims. If different speech communities have taxonomies that differ in some local area, then members of one of them can (and occasionally will) make statements that, though fully meaningful within that speech community, cannot in principle be articulated by members of the other. To bridge the gap between communities would require adding to one lexicon a kind term that overlaps, shares a referent, with one that is already in place. It is that situation which the no-overlap principle precludes.
>
> Incommensurability thus becomes a sort of untranslatability, localised to one or another area in which two lexical taxonomies differ. The differences which produce it are not any old differences, but ones that violate either the no-overlap condition ... or else a restriction on hierarchical relations ... Violations of those sorts do not bar intercommunity understanding ... But the process which permits understanding produces bilinguals, not translators, and bilingualism has a cost. The bilingual must always remember within which community discourse is occurring. The use of one taxonomy to make statements to someone who uses the other places communication at risk.[15]

So, for example, the Copernican statement 'planets travel around the Sun' cannot be translated into the Ptolemaic lexicon, because although the term 'planet' appears as a kind term in both lexicons, the two overlap without either containing all the celestial bodies contained in the other, since a fundamental change has occurred in this taxonomic category during the transition from Ptolemaic to Copernican astronomy.[16] A second

---

[14] Kuhn sometimes seemed to concede the possibility that at least in principle bilingualism can imply translatability, although he insisted that no argument has ever been offered for this claim. In particular, Kuhn insisted that Quine's argument for the indeterminacy of translation cuts no ice for translatability: Quine's radical translator is in fact a language learner, not a translator. See on this point Kuhn (1983), p. 47; (1989), p. 61; (1990), p. 300.
[15] Kuhn (1991). Reprinted in Kuhn (2000), pp. 92–3.   [16] *Ibid.*, p. 94.

important assumption underlying the untranslatability thesis is that taxonomy must be preserved for translation to be possible:

> The lexical structures employed by speakers of the languages must be the same, not only within each language but also from one language to the other. *Taxonomy must, in short, be preserved to provide both shared categories and shared relationships between them.* Where it is not, translation is impossible.[17]

On this view, translation would require a one-to-one mapping from the taxonomic categories and relationships of one lexicon to those of another lexicon at the cost of overlapping kind terms. It is this situation that the no-overlap principle precludes.

### 3.2.2 Kuhn's argument for untranslatability and Hacking's taxonomic solution to the new-world problem

I take Kuhn's argument for untranslatability to run as follows:

1. scientific lexicons are holistic
2. scientific lexicons are structured
    2.a lexical structures are taxonomic
    2.b taxonomy must be preserved for translation to be possible
    2.c any two lexicons typically display (locally) different lexical taxonomies
    2.d bridging the gap between different lexical taxonomies implies a violation either of the no-overlap principle, or of suitable restrictions on hierarchical relations

∴ scientific lexicons are untranslatable

Note that in this argument the conclusion does not necessarily follow from premises (1) and (2) alone: neither holism nor the structural character of lexicons *by themselves* implies that scientific lexicons are untranslatable. For this conclusion, the argument needs some further lemmas, namely (2.a)–(2.d). More precisely, the Archimedean platform of this argument is lemma (2.a), as the lemmas (2.b)–(2.d) spell out the implications of (2.a). But taxonomy is implied neither by premise (1) scientific lexicons are holistic, nor by premise (2) scientific lexicons are structured. A scientific lexicon can be holistic and structured without being necessarily taxonomic. Taxonomy is an extra assumption over and above premises (1) and (2) that nonetheless must be introduced for the conclusion about untranslatability

---

[17] Kuhn (1983). Reprinted in Kuhn (2000), p. 53. Emphasis added.

to follow. Kuhn forced taxonomy on lexicons *as if* there were genus–species relationships among kind terms. The prescriptive strength of the no-overlap principle itself relies on this taxonomic assumption: the principle forbids any overlapping between taxonomic categories belonging to the same contrast set, while allowing inclusive overlapping of genus–species type. We should then focus on the additional lemma (2.a) to assess the credentials of Kuhn's untranslatability thesis. But before proceeding in this direction, a ground-clearing remark about what is really at stake in Kuhn's thesis is necessary.

The untranslatability thesis challenges the rational continuity of science, the idea that scientific concepts *naturally evolve* from earlier theories to later mature ones, by replacing it with the rival view of science as a sequence of disconnected, incommensurable theories. In this scenario, bilingualism remains the only way of moving from one lexicon to another untranslatable one. However, Kuhn's view need not be the threat to the rational continuity of science that it has often appeared to be. It is in this direction that we should read the repeated attempts of philosophers of science to mitigate Kuhn's untranslatability thesis. Just to mention a few examples, Philip Kitcher has argued for the full communication of scientists across revolutions because lexicons are enrichable, i.e. they can be extended to include new terms.[18] Mary Hesse has appealed to approximate sharing or significant intersections of taxonomies to weaken the untranslatability claim.[19] Ian Hacking has proposed 'the taxonomic solution'[20] to what he calls Kuhn's problem of the new world. Since Hacking's argument is relevant to my following line of argument, I am going to present it in some detail in the rest of this section.

As mentioned above, Kuhn repeatedly claimed that scientists live or work in a new world after a scientific revolution.[21] Hacking has offered a nominalist solution to Kuhn's new-world problem, whose bottom line reads as follows: the world consists of individuals, and as such it does not change during a scientific revolution. Yet the world scientists work in and act upon is not a world of individuals but a world of kinds, and kinds typically change during a scientific revolution, where 'scientific kinds' must be distinguished from natural kinds (as they also include artifactual kinds) nor should they be confused with biological taxa familiar from natural history and systematics, despite the common philosophical origins

---

[18] Kitcher (1983).  [19] Hesse (1983).  [20] Hacking (1993).
[21] 'Though the world does not change with a change of paradigm, the scientist afterwards works in a different world ... I am convinced that we must learn to make sense of statements that at least resemble these.' Kuhn (1962), p. 121.

in the Aristotelian tradition. Having clarified that, Hacking goes on to show that 'Kuhn's taxonomies ... force scientific kinds from distinct paradigms to be untranslatable'[22] thanks to three conditions:

(1) scientific kinds are taxonomic
(2) scientific-kind taxonomies bottom out in the most specific (*infima*) species
(3) scientific terms are projectible.

Condition (3) is lexical, whereas the other two conditions concern more scientific kinds than scientific terms. The lure of translation then arises in three possible cases:

> a. *A kind overlaps a scientific kind in the new science*. Then by condition 1, the kind in the old science cannot be a kind in the new science ...
> b. *A kind subdivides a kind in the new science that has no subkinds*. By condition 2, the kind in the new science is an infima species ... and so the old names cannot be translated into any expression in the new science that denotes a scientific kind.
> c. *Although the lexicons of the old science and the new one differ taxonomically, a kind in the old science coincides with a kind in the new science*. This is the case with 'water' before and after isotopes. We should not, in my opinion, argue for untranslatability.[23]

Hacking, correctly in my opinion, recognizes that it is condition 1 that does the job for Kuhn's untranslatability thesis. He then asks, 'Is it true? Are the scientific kinds of a branch of science taxonomic? The short answer is no.'[24] We may well in fact have antitaxonomic arrangements of scientific kinds that make room for overlapping. For instance, the kind 'poison' overlaps the kinds 'vegetable' and 'mineral' (e.g. arsenic is a kind of poison and a kind of mineral; hemlock is a kind of poison and a kind of vegetable). To avoid this sort of counterexample, Hacking introduces a further condition: *scientific kinds are real Kinds*, where the notion of real Kinds is borrowed from John Stuart Mill. A real Kind is a kind characterized by a virtually 'unlimited number of properties that do not follow from the marks by which we distinguish it but that we can endeavour to find out';[25] for example, arsenic is a real Kind, but poison is not, because, do what we may, there is no way to exhaust the virtually inexhaustible number of properties of arsenic. But the question remains: 'Can we prove real Kinds are taxonomic, that no antitaxonomic examples are possible?

---

[22] Hacking (1993), p. 289.   [23] *Ibid.*, p. 295.   [24] *Ibid.*, p. 299.   [25] *Ibid.*, p. 300.

## 3.2 Reconsidering Kuhnian incommensurability

Perhaps we can if we ... define the antitaxonomic structures of science out of existence ... But again, I am unpersuaded that we should do so.'[26]

There are two main lines of argument in Hacking: the first is explicitly stated, the second is not, even if it is perhaps the more important. The first concerns antitaxonomies. The second line of argument is contained in point c of Hacking's aforementioned quotation about translatability: no matter how different lexical taxonomies can be, kinds in the old and new science may well coincide, because in the end kinds are independent of taxonomic classifications and, more generally, of any epistemic factors. The kind 'water' remained the same before and after the discovery of isotopes: notice here the surreptitious appeal to Putnam's causal theory of reference.[27] Hacking's main objection to untranslatability relies on this second line of argument, although the two arguments are inter-related: Hacking's nominalist approach to taxonomies is congenial to the Putnamian idea that kinds are metaphysically fixed independently of epistemic factors.

But no such Putnamian view is open to someone like Kuhn, who against 'those who maintain the independence of reference and meaning [and] also maintain that metaphysics is independent of epistemology' has replied that 'no view like mine ... is compatible with that separation'.[28] In his response to Hacking, Kuhn did not really address Hacking's first line of argument about antitaxonomies, to concentrate instead on the second line of argument:

> Though the solution he [Hacking] describes was never quite my own and though my own has developed substantially since the manuscript he cites was written, I take immense pleasure in his paper ... His nominalist version of my position – there are really individuals out there, and we divide them into kinds at will – does not quite face my problems. The reasons are numerous, and I mention only one here: how can the referents of terms like 'force' and 'wave front' ... be construed as individuals? I need a notion of 'kinds', including social kinds that will populate the world as well as divide up a pre-existing population. That need in turn introduces a last significant difference between me and Ian. He hopes to eliminate

---

[26] *Ibid.*, p. 301.

[27] According to Putnam (1975), the meaning of a natural kind term such as 'water' is a four-component vector with: (1) a *syntactic marker* (e.g. noun); (2) a *semantic marker* (e.g. liquid); (3) a *stereotype* (e.g. colourless, transparent, tasteless); (4) a *description of the extension* (e.g. $H_2O$). Some beliefs in the stereotype might be mistaken, and change in time and in different communities. Nonetheless, by identifying the meaning of the 'meaning' with the reference, Putnam's theory warrants referential continuity across theory-change. Following Kripke (1972), if water is $H_2O$, then it is *metaphysically necessary* that water is $H_2O$. However, as Putnam points out, metaphysical necessity is not and does not imply epistemic necessity: water is $H_2O$ even if we may well not know that water is $H_2O$. Putnam's theory is congenial to Hacking's experimental realism; see Hacking (1983), Chapter 6.

[28] Kuhn (1989). Reprinted in Kuhn (2000), p. 77, footnote 25.

all residues of a theory of meaning from my position; I do not believe that that can be done ... Both 'water₁' and 'water₂' are kind terms: the expectations they embody are therefore projectible. Some of those expectations are different, however, which results in difficulties in the region where they both apply.[29]

Kuhn argued that in the case of a polysemous term such as 'water', any attempt to resolve the tension by introducing two terms, 'water₁' and 'water₂', to designate two different meanings (water before and after isotopes) sharing nonetheless the same referents, 'is ... linguistically unsupportable'.[30] Epistemologically, 'water₁' and 'water₂' are very different kind concepts, associated with different projectible expectations. While insisting on this point, Kuhn evaded the force of Hacking's objection about antitaxonomies. Thus, lemma (2.a) remains the unquestioned Archimedean platform for Kuhn's untranslatability thesis.[31]

Despite Kuhn's dismissive attitude, Hacking's nominalist solution is, I think, on the right track in identifying and unmasking some surreptitious assumptions of Kuhn's argument. The antitaxonomy that Hacking envisages in arsenic and hemlock has the following kernel of truth: arsenic and hemlock lend themselves to be taxonomically classified under different taxonomic trees corresponding to the different properties they can be predicated of. We can pick up any two pairs of properties to build up four different taxonomic trees: arsenic can be taxonomically classified under the tree of 'minerals' as well as under the tree of 'poisons'; *mutatis mutandis* for hemlock (Fig. 3.1). Should the properties of arsenic be virtually inexhaustible, i.e. were arsenic a real Kind in Mill's sense, there

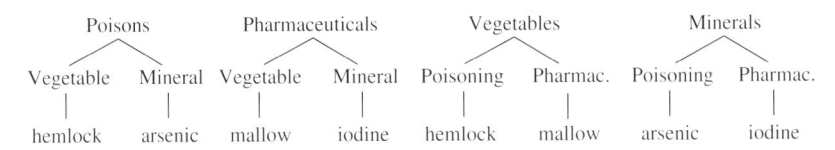

Fig. 3.1 The four taxonomic trees for the two pairs of properties poisoning/pharmaceutical and vegetable/mineral.

---

[29] Kuhn (1993), pp. 315–9.   [30] *Ibid.*, p. 318.
[31] Lexical taxonomies were the leitmotiv of Kuhn's later research. In his last and never published book Kuhn intended to investigate the neural mechanisms underlying the formation of lexical taxonomies: 'In the book I will suggest that this characteristic can be traced to, and on from, the evolution of neural mechanisms for reidentifying what Aristotle called "substances": things that, between their origin and demise, trace a lifeline through space over time. What emerges is a mental module that permits us to learn to recognise not only kinds of physical object (e.g., elements, fields, and forces), but also kinds of furniture, of government, of personality, and so on ... I shall refer to it as the lexicon, the module in which members of a speech community store the community's kind terms.' *Ibid.*, p. 315.

would be a virtually inexhaustible number of trees it could potentially be classified under. So, adding the further condition that scientific kinds are real Kinds does not eliminate antitaxonomic counterexamples.

Hacking's nominalist solution has the merit of unmasking some philosophical assumptions about lexical taxonomies underlying lemma (2.a), namely their time-honoured philosophical origins in the Aristotelian theory of predicables. Hacking's solution has itself a distinguished philosophical pedigree in Medieval nominalism, which much contributed to the demise of Aristotle's theory, as following brief overview will clarify.

### 3.2.3 Lexical taxonomies: the Aristotelian tradition and the nominalist criticism

A lexical taxonomy consists in a finite, hierarchical system where a most general kind term/kind concept progressively divides into a series of kind terms/kind concepts, until the tree bottoms out with the most specific kind terms/kind concepts, those not amenable to further division. Lexical taxonomies capturing the genus–species relationships amongst kinds originate from the Aristotelian theory of predicables. This theory was developed while looking for an adequate method for definitions: in Aristotle's view, defining an entity involves listing a series of essential properties that can be predicated of it and that divide a genus into the species to which the entity at issue belongs. For instance, 'man' is defined as 'mortal rational animal', where mortal and rational are the essential properties or specific differences that distinguish the species 'man' within the genus 'animal' from other species such as 'horse', 'ox', and so forth. Differences are the crucial tool in the division of genera into species. Aristotle's method of dividing genera into species had an enormous philosophical resonance in the Middle Ages, thanks to the work of the neo-Platonic Porphyry (third century AD), who wrote an influential introduction to Aristotle's *Categories*: the *Isagoge*. In that book, Porphyry listed five predicables: genus, species, difference, property, and accident. Accordingly, Porphyry drew a taxonomic tree for the first Aristotelian category (substance), and claimed that similar taxonomic trees can be drawn for the other Aristotelian categories taken as the most general or first genera.

However, there was a major problem with Porphyry's taxonomic tree, a problem well-known and much debated among Medieval philosophers, which I shall call the problem of symmetric hierarchy. In *De divisione* Boethius remarked how differences could be reorganized in alternative taxonomic trees: we can divide the genus of white things into liquid (milk) and hard (pearl), but we can also divide the genus of liquid things into white (milk) and black (ink). Abelard in the *Editio super Porphyrium* made a similar point: the genus 'animal' can be divided into rational/irrational (so as to distinguish 'man' from 'horse'), and then into mortal/immortal; but it can also be divided into mortal/immortal first (so as to distinguish 'man' from 'god') and then into rational/irrational. These taxonomic rearrangements highlight how the specific differences that lexical taxonomies consist of do not form an asymmetric hierarchy of interlocking properties, each embedding the subsequent ones. As Umberto Eco has illuminatingly pointed out apropos the Porphyryan tree:

> Genera and species are linguistic ghosts that cover the real nature of the tree and of the universe it represents: a world of pure differentiae ... Since differentiae do not contain each other, the classical Porphyrian tree ... *is no longer a hierarchical and ordered structure* ... In a tree composed with pure differences, these differences can be rearranged *according to the description under which a given subject is considered*.[32]

Given any pair of differences, there is no warrant that the first contains the second any more than the second contains the first. Going back to Hacking's antitaxonomy for arsenic and hemlock, the four taxonomic trees for the pairs of differences mineral/vegetable and poisoning/pharmaceutical originate exactly from this problem (Fig. 3.1). As Abelard was well aware, differences are only *nomina*, names, whose taxonomic order can be reversed. It is illusory to think that they form a chain of interlocking properties, from the upper to the lower ones. This illusion is a residue of Porphyry's neo-Platonic leanings, a residue that lexical taxonomies still seem to carry.

The aim of this brief overview is to suggest that it is the enduring philosophical legacy of the Aristotelian tradition that is surreptitiously at work in Kuhn's discussion of lexical taxonomies, while Hacking's nominalist solution is on the same conceptual path of Medieval nominalists in the criticism of the Porphyryan tree. This is both good and bad news for Kuhn. It is good news because Kuhn can legitimately respond that the nominalist solution has never been his own and does not quite face his problems. But it is also bad news as it unveils the surreptitious Aristotelian/neo-Platonic gloss that Kuhn needs to put on lexical taxonomies for the

---

[32] Eco (1984), pp. 65–6.

untranslatability argument to go through, and for the new-world problem to arise.[33] I want to latch my criticism of Kuhn onto this last point.

### 3.2.4 How should we read lexical taxonomies? A Kantian reading

As we saw in Section 3.2.2, Kuhn's argument for untranslatability needs lemma (2.a), which identifies lexical structures with taxonomic structures, to go through. Once we question this lemma, or better once we question the specific reading that Kuhn gave of lexical taxonomies, the conclusion about untranslatability needs to be reconsidered. I am not going to argue for translatability: although lemma (2.b) about translation being taxonomy-preserving seems too strong a requirement, I am not going to dispute it.[34] There may still be no translation, if by translation we mean a one-to-one mapping between taxonomic categories of different lexicons. Yet, if lexical structures are not necessarily taxonomic in the strong sense that Kuhn conveyed, as I am going to argue, the Archimedean platform of Kuhn's argument is undermined. A legitimate question to ask then is how we should read the taxonomic arrangements of genus–species type among kind terms/kind concepts of a lexicon.

Hacking opts for a weak nominalist reading, one that can dissolve (more than solve) the new-world problem. But nominalism is at odds with Kuhn's overall view, and not only for the reason that Kuhn explicitly mentioned in his response, namely that it is hard to conceive how to construe in a nominalist way terms such as 'force' and 'wave front'. I think there is a deeper clash between nominalism as the view that the world consists of individuals and we carve scientific kinds 'at will', on the one side, and Kuhn's self-declared 'post-Darwinian Kantianism' on the other side. As mentioned in Section 3.2.1, in his later writings Kuhn came to present his position as a form of Kantianism, where the lexical categories of a scientific lexicon would supply the preconditions of possible

---

[33] For the new-world problem, see Footnote 21.
[34] For instance, Kuhn's claim that it is impossible to translate into French the English statement 'the cat sat on the mat' because of the different lexical taxonomies for floor coverings in French and in English (see Kuhn 1991, reprinted in Kuhn 2000, p. 93) is obviously true, but also unilluminating. There is an embarrassment of riches of similar lexical situations in any translation from one language to another. And there are plenty of linguistic strategies to render words that do not have any direct equivalent in another language. Skilled translators know very well how to do this job. Claiming that translation must be taxonomy-preserving seems too narrow a criterion, and almost begs the question for Kuhn's thesis.

experience and yet, by contrast with Kant, lexical categories can and do change after a scientific revolution. Kuhn wanted to maintain a Kantian position, albeit a dynamic one: no wonder he could not recognize as his own the nominalist solution, according to which we can carve and reshuffle lexical categories at will.

On the other side, I think Kuhn gave a strong (Porphyryan-like) reading of lexical categories. In the overall strategy of his argument for untranslatability, lexical taxonomies were entrusted with the crucial role of opening up a world for the speakers of a certain lexicon, to the extent that if a fundamental change occurs in some taxonomic categories during a scientific revolution, speakers of the old and the new lexicon would not only be unable to communicate unproblematically, but they would in fact end up 'living in different worlds'. Locally different lexical taxonomies give access to different worlds. Kuhn's problem of the new world is inherently related to lexical taxonomies and to the role that they play via lemma (2.a) in Kuhn's argument for untranslatability.

There is a surprising and generally overlooked asymmetry, I think, between Kuhn the historian and Kuhn the philosopher of science. As a historian, Kuhn made an unprecedented contribution to the reappraisal of the long-lasting and ceaseless process of conceptual revision and theoretical refinement that typically takes place during periods of crisis and scientific revolution, as his excellent monographs on the Copernican revolution and on black-body theory testify.[35] However, as a philosopher of science, Kuhn emphasized instead the discontinuities and gaps between paradigms: incommensurability, both in its earlier and later (mitigated) version, has remained Kuhn's philosophical manifesto.

For the argument for untranslatability to go through, Kuhn the philosopher had to give a strong (Porphyryan-like) gloss to lexical taxonomies, which did the job for incommensurability as much as it was – I think – at odds with the ceaseless process of conceptual revision that Kuhn the historian on the other hand so well described. By 'Porphyryan-like gloss' I mean that Kuhn regarded lexical taxonomies as displaying a 'world', as fixing an order of things in nature, hence the claim that scientists employing different lexical taxonomies (before and after a scientific revolution) live and work in different worlds.

In the rest of this section I foreshadow an alternative Kantian reading of lexical taxonomies. On this reading, we can still maintain lexical taxonomies without dissolving them into clusters of names. But, on the other

---

[35] Kuhn (1957), (1978).

## 3.2 Reconsidering Kuhnian incommensurability

hand, we no longer give them a strong gloss. A Kantian reading is equidistant from the weak nominalist reading and from the strong Aristotelian/neo-Platonic reading. There are two main reasons for preferring it. It does justice to Kuhn's self-declared 'post-Darwinian Kantianism', and it is in better agreement with the view that Kuhn the historian delivered about scientific revolutions. I clarify the first point in this section, and the second in the following one. But let me first clarify how a Kantian reading of lexical taxonomies would look.

As we saw in Section 1.4.1, Kant regarded the division of genera into species as the expression of the three logical principles of homogeneity, specification, and continuity through which the regulative principle of systematic unity is articulated. In this view, we pursue homogeneity by subsuming empirical concepts under higher-order concepts, i.e. more abstract and more encompassing genera, as much as we pursue variety by expanding and specifying lower-order concepts, i.e. the possible species and subspecies within the same genus. Continuity is necessarily presupposed, as it is only under the assumption that there is a *continuum formarum* that the genus–species relationships among empirical concepts are amenable to being modified by introducing new intermediate concepts either in the upper or in the lower nodes of the taxonomic division. However, as Kant pointed out, this continuity is only an idea as much as the division of genera into species that the principles of homogeneity and specification give rise to is an idea. Taxonomic genus–species relationships among concepts must not be regarded as objectively grounded in nature or as mirroring an order of nature. Rather, they are functional to systematic unity as a regulative principle[36] that helps us to find our way

---

[36] As mentioned in Section 1.4.1, Kant did not regard systematic unity just as a logical, but rather as a transcendental principle, because for systematic unity to be conceivable as a goal (albeit an unattainable one in practice), we must presuppose that such an order is grounded in nature. However, we cannot give a transcendental deduction for this, which remains a regulative (not a constitutive) principle. See Kant (1781), English translation (1997) A668/B696. In a different way, Kant restated the same point in the First Introduction of the *Critique of the Power of Judgment*, where he notices that given the great variety and diversity of natural forms, we need to presuppose in nature a certain uniformity that we can grasp, and 'this presupposition, as an a priori principle of the power of judgment, must precede all comparison', for 'without [this] presupposition we could not hope to find our way in a labyrinth of the multiplicity of possible empirical particular laws'. This a priori, transcendental principle of the reflecting power of judgment is merely 'a principle for the logical use of the power of judgment … for the sake of regarding nature a priori as qualified for a *logical system* of its multiplicity under empirical laws', where 'the logical form of a system consists merely in the division of given general concepts … by means of which one thinks the particular (here the empirical) with its variety as contained under the general' Kant (1790), English translation (2000), First Introduction, Section V, 20: 214–20: 215. In other words, the taxonomic *classification* of the empirical manifold is

through the empirical manifold and to recognize empirical regularities as necessary laws.

Thus, on a Kantian reading, lexical taxonomies can be regarded as accomplishing a merely regulative task. Scientists put forward lexical taxonomies while striving to organize the empirical manifold into a coherent system of scientific knowledge. Different lexical taxonomies are simply the different expressions of one and the same regulative demand that enjoins us to pursue systematicity, and that manifests itself through different systems of scientific knowledge. In the attempt to make sense of anomalous phenomena, scientists end up resystematizing scientific knowledge by reshaping and reorganising some concepts; introducing new intermediary concepts to reconcile anomalous evidence with old and increasingly inadequate concepts; redefining the taxonomic relationships among them. This ceaseless process, through which scientific concepts cycle back and forth, get redefined until they finally acquire a new meaning, is the manifestation of the open-ended regulative task of pursuing systematicity, of 'projecting' an order upon nature.

On this view, the new lexical taxonomy that emerges after a scientific revolution should not be understood as literally opening up a 'new world', as disclosing a new order of things – as Kuhn suggested by giving a strong (Porphyryan-like) gloss on lexical taxonomies – since it is only the provisional and tentative expression of the never fully attainable goal of systematization. On the other side, with this Kantian reading, we can give a new twist to Kuhn's claim that after a scientific revolution we live in a new world, if by 'world' we now mean the phenomenal world we have epistemic access to within the boundaries of our current system of knowledge. And since from a dynamically Kantian perspective, this is the only world we are entitled to investigate and to have knowledge of, when it changes, after a scientific revolution with the rise of a new lexical taxonomy that by encompassing previously recalcitrant phenomena 'projects' a new order

---

here too presented as satisfying a regulative requirement (imposed by the faculty of reason, and carried out by the faculty of reflecting judgment), namely the requirement of thinking nature as a logical system, whereby particular lower-order concepts (species) are thought of as contained and subsumed under more abstract, higher-order concepts (genera). Kant identified the transcendental principle of the reflecting judgment as the 'formal purposiveness of nature', which we must necessarily presuppose in order to have 'an order of nature in accordance with empirical laws', and hence for systematic unity to be at least in principle attainable as a goal, without however forgetting that this is only a 'subjective principle (maxim) of the power of judgment' that cannot be proved and for which no transcendental deduction is available (see on this point also Sections IV and V of the published Introduction of the *Critique of the Power of Judgment*, 5: 183–5: 186).

upon nature, we can legitimately say – in a somehow liberal sense – that we 'live in a different world'.

Thus, a mild Kantian reading of lexical taxonomies allows us to reformulate the new-world problem in a way that does not license incommensurability and, on the other side, vindicates Kuhn's 'post-Darwinian Kantianism'. It achieves what Kuhn merely declared as a programme. Indeed, Kuhn's strong reading of lexical taxonomies seems based on a 'taxonomic fallacy', so to speak: despite the declared 'post-Darwinian Kantianism', Kuhn conflated the regulative *as if* with *is*. He took the genus–species relationships among kind concepts/kind terms not as fulfilling a merely regulative demand in the Kantian sense (i.e. *as if* nature were ordered according to those taxonomic relationships), but rather as fixing an order of things in nature in the Aristotelian/neo-Platonic sense (i.e. nature *is* so ordered). Only by taking lexical taxonomies as satisfying a regulative demand, can we do justice to a truly dynamic Kantianism like the one Kuhn was willing to embrace. Furthermore, only in this way can we reconcile Kuhn the philosopher with Kuhn the historian, and give history its due, as I now turn to clarify in the light of the historical episode reconstructed in Chapter 2.

### 3.2.5 Reintroducing history in scientific lexicons: a lesson from the crisis of the old quantum theory

If Kuhn's view is correct, the passage from the old quantum theory to the new quantum theory reconstructed in Chapter 2 should be read as the shift from a lexical taxonomy to a different (untranslatable) one. The scientific revolution around 1921–5 sanctioned the end of old taxonomic categories, such as the atomic core angular momentum, and the rise of new ones, namely the electron's spin angular momentum. Taxonomic relationships among concepts were also redefined: the problematic half-integral values – originally assigned to the atomic core in Heisenberg's sharing principle or to a mysterious quantity $d$ in Bohr's – were finally attributed to the electron's spin. If lexical structures were taxonomic in the strong sense that Kuhn conveyed, the story told in Chapter 2 would provide us with an example of incommensurability as untranslatability. Consider, for instance, the physical concept of momentum. In the old quantum theory lexicon, this concept can be regarded as the upper genus of the taxonomic tree in Fig. 3.2. The lower nodes of this taxonomic tree are very different from those of the taxonomic tree in Fig. 3.3, which portrays the new quantum theory lexicon where the

## 3 From the old quantum theory to the new quantum theory

```
                        Momentum
                       /        \
                  Angular      Non-angular
                 /      \
        Orbital (electron)    Non-orbital (atomic core)
        /        \              /            \
   Integral   Half-integral   Integral    Half-integral
   [Bohr]   [Heisenberg's    [Bohr]    [Heisenberg's
            sharing principle]          sharing principle]
```

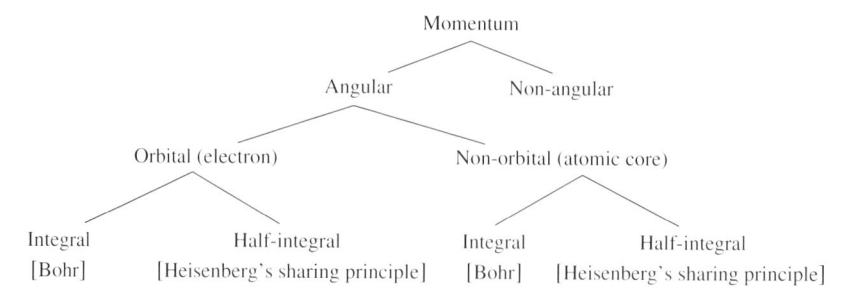

Fig. 3.2 Lexical taxonomy for the kind term 'momentum' in the old quantum theory, with in round brackets the entities at issue and in squared brackets the rival theoretical proposals.

difference orbital/non-orbital divides the species 'angular momentum', and these subspecies can be further divided according to the differences integral valued/half-integral valued, where finally the difference two-valued (*zweideutig*) referred to the half-integral non-orbital angular momentum is ascribed to the electron's spin. If we put a strong (Porphyryan-like) gloss on these taxonomic arrangements, we may say that the fundamental change of some taxonomic categories in the passage from the old to the new quantum theory lexicon has made the two lexicons untranslatable.

But it is exactly this strong gloss that I resist subscribing to, and for good historical reasons, beside the ones discussed in the preceding section and concerning more directly Kuhn's *soi-disant* dynamic Kantianism. The revolutionary transition from the old to the new quantum theory lexicon was far more articulated than an allegedly abrupt shift from the lexical taxonomy of Fig. 3.2 to that of Fig. 3.3. Looking back at the historical episode reconstructed in Chapter 2, it is evident that this revolutionary transition passed instead through a series of intermediate steps that gradually led to the conceptual revision of the old taxonomic categories, and to their final substitution with new ones. Pauli himself worked for a few years on the atomic core model; he introduced the concept of a *zweideutig* angular momentum in September 1923 to describe the electron's angular momentum no less than the core's and the atom's angular momenta, as the 'true' angular momenta at work in Landé's *g* formula (see Section 2.2.3). Inspired by Pauli's suggestion, in May 1924 Landé and Heisenberg reinterpreted the *zweideutig* angular momentum of the atom in terms of their branching rule [*Verzweigungsregel*]: the total angular momentum of the atom seemed to branch into two possible half-integral values ($j \pm 1/2$) during

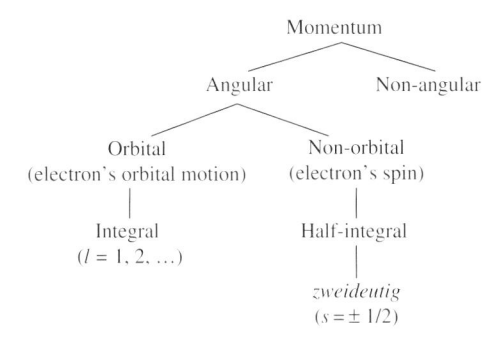

## 3.2 Reconsidering Kuhnian incommensurability   99

```
                        Momentum
                       /        \
                  Angular      Non-angular
                  /     \
            Orbital    Non-orbital
     (electron's orbital motion)  (electron's spin)
            |                   |
         Integral          Half-integral
       (l = 1, 2, ...)           |
                              zweideutig
                             (s = ± 1/2)
```

Fig. 3.3 Lexical taxonomy for the kind term 'momentum' in the new quantum theory (after 1925), where the *zweideutig* half-integral values for the non-orbital angular momentum are ascribed to the electron's spin.

the building-up process. A month later, in June 1924, the branching rule was converted into Heisenberg's new quantum principle, where no mention of the branching process could be found (Section 2.2.3), and the *zweideutig j* now came to fix the interval over which the classical Hamiltonian for the anomalous Zeeman effect should be integrated. In the meantime, using the rule of the permanence of the *g* sums, Pauli could avail himself of the spectroscopic data on the Paschen–Back effect to shed light on the anomalous Zeeman effect. After Landé's discovery of the null result for the relativistic correction factor in thallium, in November 1924 Pauli finally arrived at the conclusion that the valence electron alone was responsible for the observed spectroscopic anomalies: in a non-mechanical way, the electron runs about in two (*zweideutig*) different states with the same orbital angular momentum. And in 1925 Uhlenbeck and Goudsmit interpreted the electron's *Zweideutigkeit* as the two-valued intrinsic angular momentum of a spinning electron.

It is this complex stepwise process, through which scientific concepts cycle back and forth to be gradually redefined until they acquire new meanings, that becomes irredeemably lost in the allegedly abrupt interlexical taxonomy shift that Kuhn's strong reading delivers. This is a process that Kuhn the historian knew well, but that Kuhn the philosopher seemed willing to sacrifice in the name of the philosophical agenda about incommensurability. With a view to remedying this asymmetry between Kuhn the historian and Kuhn the philosopher, the alternative Kantian reading of lexical taxonomies foreshadowed in the previous section can be fruitfully applied. On this alternative reading, lexical taxonomies would be the always provisional expression of a regulative demand that enjoins

100    3 *From the old quantum theory to the new quantum theory*

```
                              Momentum
                             /        \
                        Angular      Non-angular
                       /       \
                    Total     Non-total
                   /     \
         Half-integral   Integral
    (riddle of statistical weights)
           /        \
     zweideutig   Not zweideutig
  (Landé–Heisenberg branching rule)
```

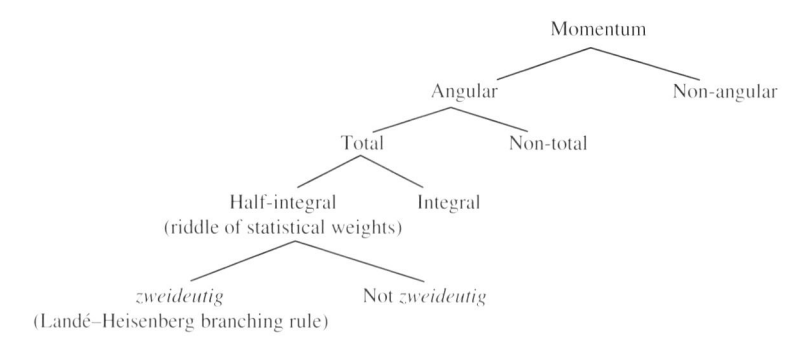

Fig. 3.2* Lexical taxonomy intermediate between 3.2 and 3.3, reflecting the historical process through which half-integral values were first assigned to the total angular momentum of the atom, whence the riddle of statistical weights, and the Landé–Heisenberg branching rule.

scientists to pursue the open-ended goal of systematization, a systematization of the body of scientific knowledge able to incorporate an increasing number of anomalous phenomena by reshaping and reorganizing scientific concepts, introducing intermediary ones, until they eventually result in brand new concepts.

We can accordingly portray the revolutionary transition reconstructed in Chapter 2 as mediated by a continuous series of lexical taxonomies, intermediate between that of Fig. 3.2 and that of Fig. 3.3, and obtained by interpolating or rearranging the lower nodes of the taxonomy of Fig. 3.2, while striving for a better systematization of the body of knowledge able to fit puzzling phenomena. To put the matter in a somehow pictorial way, we should introduce a taxonomic tree, intermediate between the tree of Fig. 3.2 and that of Fig. 3.3 (shown as Fig. 3.2*) and portraying the theoretical step through which physicists discovered that the total angular momentum of an atom was half-integral valued, where this puzzling half-integral value was then assigned a ($\pm$) sign with the Landé–Heisenberg branching rule.

But the introduction of this intermediate tree does not exhaust the complexity of the historical transition around 1921–5. It still leaves out a previous important step, namely the one concerning Pauli's original introduction of the concept of a *zweideutig* angular momentum for the core, the electron, as well as the atom, from which the Landé–Heisenberg branching rule followed. So, the taxonomic tree of Fig. 3.2* can in turn be regarded as stemming from this other taxonomic tree (shown as Fig. 3.2**), intermediate between 3.2 and 3.2*.

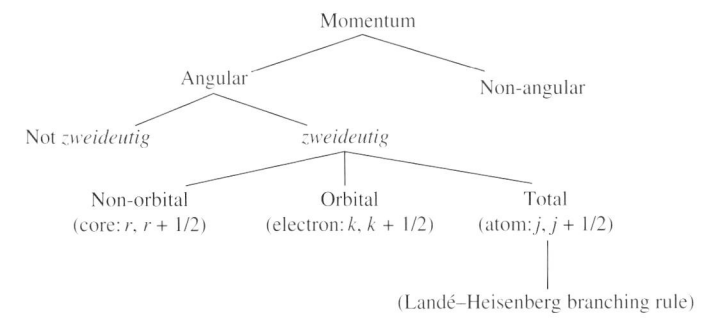

## 3.2 Reconsidering Kuhnian incommensurability

```
                        Momentum
                       /        \
              Angular              Non-angular
             /      \
     Not zweideutig   zweideutig
                     /    |    \
            Non-orbital  Orbital   Total
            (core: r, r + 1/2)  (electron: k, k + 1/2)  (atom: j, j + 1/2)
                                                          |
                                              (Landé–Heisenberg branching rule)
```

Fig. 3.2** Lexical taxonomy intermediate between 3.2 and 3.2*, reflecting Pauli's original theoretical step of regarding the angular momentum of the core, of the electron, and of the atom as twofold (*zweideutig*) in Landé's formula for *g* factors, which inspired the Landé–Heisenberg branching rule.

Recalling the problem of symmetric hierarchy in Section 3.2.3, we may say that the difference *zweideutig*/not *zweideutig* is *not* contained (necessarily) into the difference integral/half-integral, any more than integral/half-integral is contained (necessarily) in the difference orbital/non-orbital. These differences cut across the lexical taxonomies above, and they do not form an asymmetric hierarchy of interlocking properties.

Notice, however, that this does not imply that their order can be re-shuffled arbitrarily, or that any of these concepts can be carved at will, as a nominalist reading would suggest. There are instead specific experimental and theoretical constraints on the possible taxonomic rearrangements. The sequence of Figs. 3.2, 3.2**, 3.2*, and 3.3 respectively reflects the subsequent redescriptions that the inter-related concepts of 'non-orbital angular momentum', 'half-integral values', and '*Zweideutigkeit*' underwent through the experimental and theoretical developments of September 1923, May 1924, and 1925, so that we can regard the lexical taxonomy of Fig. 3.2 as *naturally evolving* into that of Fig. 3.3, via the stepwise process portrayed by the intermediate taxonomies of Fig. 3.2** and Fig. 3.2*.

The demise of the old category of the atomic core and the rise of the new category of the electron's *Zweideutigkeit* was the result of a process of gradual theoretical shift and conceptual revision involving the experimental refinement of some parameter values (e.g. the measurement of the relativistic correction factor in chemical elements with high atomic number); the redescription of some phenomenological laws (e.g. the branching rule transformed and embedded into Heisenberg's new quantum principle), as well as the reinterpretation of some fundamental theorems such as Larmor's theorem, whose violation was initially ascribed to a core magnetic anomaly. It is via this series of experimental and theoretical constraints that the old taxonomic categories came to be progressively modified in the endeavour to achieve a better systematization of the scientific knowledge available and to reconcile it with an increasing number of anomalous phenomena, until those categories became eventually unrecognizable. When a scientific lexicon becomes unrecognizable, this is the signal that the evolution of scientific concepts has reached an irreversible turning point. It is at this point that the transition from the old to the new lexicon becomes not only intelligible, but feasible.

The crisis of the old quantum theory, which I have reconstructed in Chapter 2, nicely illustrates and supports an intermediate Kantian reading of lexical taxonomies, alternative both to Kuhn's strong reading and to the weak nominalist reading. This Kantian reading has the merit of reintroducing history in scientific lexicons. Lexicons do have a history: their categories and taxonomic relationships are subject to a continuous evolution. Behind what seems a prima facie petrified taxonomic arrangement, there is instead a state of constant ferment that gradually and irreversibly reshapes kind concepts and redefines their scope of applicability, in the never-ending pursuit of systematization. In this respect, the Kantian reading that I am suggesting goes along the lines of Thomas Nickles in urging, against Kuhn, a reinterpretation of revolutionary transitions:

> Research is recursive: it cycles back and refines such results, investing them with more meaning, as new developments occur on various fronts. Therefore, we should expect Kuhnian exemplars themselves (and the problems that they solve) to have a history ... [Kuhn] could have brought his evolutionary account of 'progress through revolutions' home to normal science ... Fitting your current problem to available exemplars is a two-way street, since matching is a symmetric relation. It is a question of mutual fit, after all. You not only deform your problem so that it better fits an exemplar; you may also find a way to deform the exemplar in order to achieve a better fit with your problem. Successful work that stretches the exemplar in this way may result in a gradual reinterpretation of the exemplar itself.[37]

Equipped as we are with a criticism of Kuhnian incommensurability, we can now give a positive account of the prospective intelligibility of the revolutionary transition that the electron's *Zweideutigkeit* brought with it.

---

[37] Nickles (2003), pp. 155 and 169.

## 3.3 The prospective intelligibility of the revolutionary transition from the atomic core model to the electron's *Zweideutigkeit*

As mentioned in Section 3.2.1, Kuhn's untranslatability thesis arises at two different levels: at the level of kind concepts, and the level of the nomic generalizations associated with them. Whenever two lexicons display (locally) different taxonomies, it is not only the taxonomic categories that cannot be translated, but also the nomic generalizations associated them. Investigating how brand new nomic generalizations may arise and become live options for practitioners of the older lexicon then becomes all the more relevant to a defence of the rational continuity of science and of the prospective intelligibility of scientific revolutions. To this end, we need to show that the process that leads to brand new nomic generalizations – inconceivable and even ineffable in the older lexicon – displays in fact some continuity with the old lexicon.

I am going to argue that in the case of Pauli's exclusion rule rational continuity should be looked for in the gradual stretching process that took place between the old quantum theory on the one side, and spectroscopic anomalies on the other side, and which finally led to the demise of the atomic core and to the introduction of the electron's *Zweideutigkeit*. But how can we more precisely characterize this stretching process? Is there any methodological framework able to capture it? In what follows I am going to argue that the methodological framework that captures this stretching process is demonstrative induction.

### 3.3.1 The electron's *Zweideutigkeit* and Pauli's exclusion rule as the conclusions of two nested demonstrative inductions

Demonstrative induction is a scientific methodology that goes back to Newton's method of deduction from phenomena. In the General Scholium of *Principia*, after recalling the difficulties with Descartes's hypothesis of vortices, Newton famously defended the law of universal gravitation:

> hitherto I have not been able to discover the cause of these properties of gravity from phenomena, and I feign no hypotheses; for whatever is not deduced from the phenomena is to be called an hypothesis; and hypotheses, whether metaphysical or physical ... have no place in experimental philosophy.[38]

---

[38] Newton (1687); English translation (1803), p. 314.

In the Rules of Reasoning in Philosophy, Newton further recommended that

> In experimental philosophy we are to look upon propositions collected by general induction from phenomena as accurately or very nearly true, notwithstanding any contrary hypotheses that may be imagined, till such time as other phenomena occur, by which they may either be made more accurate, or liable to exceptions. This rule we must follow, that the argument of induction may not be evaded by hypotheses.[39]

Not only should we not feign hypotheses, but we should also not evade the conclusions drawn from inductively generalized phenomena by conjecturing possibly contrary hypotheses. Against the proliferation of scientific hypotheses, Newton recommended the method of deducing theories from phenomena, as he systematically deployed it both in the *Principia* and in other works, from the *Optics* to the *System of the World*.[40]

Present-day defenders of demonstrative induction[41] are on the same methodological path as Newton. That what Newton portrayed as a deduction from phenomena can be more accurately described as a non-ampliative inference from one or more so-called phenomenal premises featuring inductive generalizations of empirical regularities, and from one or more so-called major premises featuring relevant theoretical assumptions taken from the scientific background knowledge. A theoretical conclusion is derived from these two different kinds of premises.[42] The non-ampliative[43] nature of the inference does not free demonstrative induction of an inductive risk, which is not located in the inferential rule as in any inductive methodology, but rather in the phenomenal premises. These are low-level generalizations inductively obtained from experimental data. This inductive risk does not however spread through the skeleton of the inference; nor does it affect the conclusion that is deemed to be (nearly) certain given the (near) certainty of the theoretical assumptions of the scientific background

---

[39] *Ibid.*, Book III, English translation (1803), p. 162.
[40] See Harper (1990); Harper and Smith (1995).
[41] Dorling (1973), (1974), (1991); Norton (1993), (1994), (1995).
[42] As an example, consider John Norton's (1994) analysis of the discovery of quantum discontinuity, deduced from a phenomenal premise about black-body radiation, and a major premise about Boltzmann's statistical mechanics (i.e. a system of black-body radiation consists of a large number of component subsystems exchanging energy in dynamical equilibrium).
[43] Ampliative inferences are those whose conclusion expands the information already contained in the premises (e.g. inductive inferences, where the conclusion is a universal generalization obtained from a finite number of positive instances listed in the premises). Non-ampliative inferences, on the other hand, are those whose conclusion does not add any information to the one already contained in the premises (e.g. deductive inferences).

Table 3.1

**Phenomenal premises**: *Spectroscopic anomalies in the early 1920s*

- (i) X-ray relativistic doublets
- (ii) Alkali optical doublets (together with (i): 'doublet riddle')
- (iii) Decrease of the number of states by half-unit in the building-up process ('riddle of statistical weights')
- (iv) Higher multiplets
- (v) Term splitting in a weak magnetic field (anomalous Zeeman effect)
- (vi) Term splitting in a strong magnetic field (Paschen–Back effect)

**Major premises**: *Theoretical assumptions of the old quantum theory*

- (a) Bohr's building-up principle
- (b) Bohr–Sommerfeld quantum conditions
- (c) Sommerfeld's space quantization
- (d) Larmor's theorem
- (e) Landé $g$ factors
- (f) Sommerfeld's relativistic explanation of the X-ray fine structure

∴ Electron's *Zweideutigkeit*

---

knowledge, which tend to remain fairly stable until a scientific revolution occurs.[44] Thus, demonstrative induction combines the non-ampliative nature typical of deductive inferences with the inductive nature of the phenomenal premises; hence the apparent oxymoron of the name.

Although demonstrative induction is primarily a method of scientific discovery (along Newton's original lines), more recently the philosophical debate has concentrated instead on the pros and cons of demonstrative induction as a method of scientific confirmation. In what follows, I consider demonstrative induction as a method of scientific discovery without discussing its possible implications for scientific confirmation.[45]

The electron's *Zweideutigkeit* can be regarded as the conclusion of a demonstrative induction featuring spectroscopic anomalies as phenomenal premises, and some relevant theoretical assumptions of the old quantum theory as major premises (Table 3.1).

---

[44] John Norton (1994), in particular, has advocated demonstrative induction as a method delivering certainty in science.
[45] I have addressed this further issue in relation to the rationality of theory-choice and the problem of underdetermination of theory by evidence, in Massimi (2004b).

The electron's *Zweideutigkeit* was not plucked out of the air as a bold conjecture. It rather flowed from spectroscopic anomalies with the help of the theoretical assumptions (a)–(f), i.e. it came out in a non-ampliative way from the interplay between the old quantum theory and anomalous phenomena as Table 3.2 highlights.

The above inference is non-ampliative: the electron's *Zweideutigkeit* was somehow already 'there' in the anomalous spectroscopic phenomena that could not be accounted for otherwise (e.g. by the atomic core model with its alleged core–electron coupling) on pain of violating some important theoretical assumptions. On the other side, induction enters this inference via the phenomenal premises, which are low-level generalizations inductively drawn from a huge amount of data about alkali spectra, X-ray spectra, Paschen–Back splitting, and so forth.

I want to suggest that this phenomenal–theoretical structure mirrors the two poles between which a two-way stretching process took place. In other words, the stretching process between those anomalous spectroscopic phenomena and those theoretical assumptions of the old quantum theory, which led to the electron's *Zweideutigkeit* as the conclusion of the demonstrative induction of Table 3.1, is the very same stretching process that ultimately led to the crisis of the old quantum theory. The prospective intelligibility of this revolutionary transition should be looked for in this process. The new quantum theory (at least as far as the electron's spin is concerned) stemmed naturally from the progressive stretching of the old quantum theory under the effect of an increasing number of spectroscopic anomalies. This stretching process has an upward path (from anomalous phenomena to theory) and a downward path (from theory to phenomena).

The first natural reaction was to reconcile anomalous phenomena with the old quantum theory by introducing suitable auxiliary assumptions. Bohr's *Zwang*, Heisenberg's *sharing principle*, the Landé–Heisenberg *branching rule*, and Heisenberg's second core model, in different ways and forms, all tried to achieve this goal, and they all failed. The new auxiliary assumptions created only further tension within the Bohr–Sommerfeld atomic theory. For instance, Heisenberg's *sharing principle* could save the anomalous Zeeman effect and the Paschen–Back effect at the cost of violating the building-up principle, the Bohr–Sommerfeld quantum conditions, Larmor's theorem, and Sommerfeld's space quantization. Heisenberg's second core model could maintain some of these main theoretical assumptions on pain of leaving unexplained the anomalous Zeeman effect. The upward path from the anomalies to the old quantum theory (via the introduction of suitable auxiliary assumptions able to accommodate recalcitrant evidence) did not resolve the tension.

The downward path from the old quantum theory to the anomalies turned out to be more fruitful. Only by stretching and progressively deforming the theory itself so as to better fit the anomalous phenomena, could the tension finally be solved. This involved (1) the introduction of new phenomenological laws such as Landé's formula for $g$ factors, and Pauli's permanence of $g$ sums; (2) the reinterpretation of well-entrenched, yet violated theoretical assumptions, such as Larmor's theorem and Sommerfeld's space quantization. As far as Sommerfeld's space quantization is concerned, its experimental confirmation due to the Stern–Gerlach experiment in 1921 became meaningful only after the introduction of the electron's spin.[46] On the other side, the violation of Larmor's theorem did not redound to the discredit of the theorem itself: Larmor's theorem retained its validity, and its failure was finally ascribed to the anomalous spin magnetic moment, responsible for the non-strict applicability of the theorem in the spectroscopic context involving the spin–orbit coupling.

While the scope of applicability of some well-entrenched assumptions came to be restricted, new phenomenological laws were also introduced. Pauli's permanence of $g$ sums established that the $g$ sums were invariant in the transition from strong to weak fields. Hence, given the known term values of the Paschen–Back effect, Pauli could proceed to calculate the corresponding term values for the anomalous Zeeman effect. But, most importantly, it was thanks to this same phenomenological law that Pauli could associate each electron with the four quantum numbers $n_{k_1,m_1,m_2}$ he had previously introduced in his analysis of the Paschen–Back effect. In so doing, he could reinterpret Stoner's rule (see Section 2.2.4): the number of states an electron can take up in a strong magnetic field coincided with the number of states an electron can take up in the $n$-th period of the periodic table, because the $g$ sums are permanent in the transition from strong fields (Paschen–Back effect) to zero fields (normal atoms following Rydberg's rule for electronic distribution). Once this crucial step had been taken, the door was open to Pauli's exclusion rule.

Thus, the exclusion rule was itself derived from inductively generalized evidence about term splitting in strong fields via Stoner's rule suitably reinterpreted in the light of the permanence of $g$ sums, via Rydberg's 'cabalistic' formula, and the electron's *Zweideutigkeit* itself. The exclusion rule can be regarded as the conclusion of the demonstrative induction of

---

[46] See Chapter 2, Footnote 8.

Table 3.2

The points (i)–(vi) below indicate the salient phenomena (the phenomenal premises are underlined). The salient theoretical assumptions (major premises (a)–(f)) are in bold character. The schema highlights the interplay of phenomena and theoretical assumptions in the derivation of the electron's *Zweideutigkeit* in Table 3.1. In order to clarify the non-ampliative nature of the inference (i.e., the electron's spin was already 'there' in the phenomena, once some important theoretical assumptions had been taken in due account), I make some theoretical comments in squared brackets.

(i) The observed doublet fine structure in X-ray spectra was given by the formula[a]

$$T = \frac{R(Z-\sigma)^2}{n^2} + \frac{R\alpha^2(Z-s)^4}{n^4}\left(\frac{n}{k} - \frac{3}{4}\right) + \frac{R\alpha^4(Z-s)^6}{n^6}\left(\frac{1n^3}{4k^3} + \frac{3n^2}{4k^2} - \frac{3n}{2k} + \frac{5}{8}\right) + \cdots$$

where $k = 1, 2, 3, \ldots$ for s, p, d ... electrons, respectively (**b. Bohr–Sommerfeld quantum conditions**). The fine structure corrections (second term, third term, ...) were attributed to the relativistic increase of the mass of the electrons (**f. Sommerfeld's relativistic explanation of the X-ray fine structure**). [They are actually jointly due to relativity and to what we now call the electron *spin–orbit coupling* (hence the name *spin-relativity doublets*)].[b]

(ii) The observed alkali doublet fine structure required the introduction of half-integral quantum numbers. These half-integral values were originally attributed either to a mysterious quantity $d = 1/2$ (Bohr); or to the electron sharing its angular momentum with the core – with then (**b. Bohr–Sommerfeld quantum conditions** – violating) $k = 1/2, 2/3, \ldots$ (Heisenberg's *sharing principle*); or to $j$ branching into $j \pm 1/2$ (**a. Building-up principle** – violating Heisenberg's *branching rule*). No unification between X-ray doublets and alkali doublets (doublet riddle) [The half-integral values are actually due to the electron's spin contribution in the spin–orbit coupling: $\mathbf{j} = \mathbf{l} \pm \mathbf{s}$ where $\mathbf{l}$ is the orbital angular momentum of the valence electron and $\mathbf{s}$ its spin with (neither **b. Bohr–Sommerfeld quantum conditions** nor **a. Building-up principle** – violating) half-integral values $\pm 1/2$. In the semi-classical model of the spinning electron, the electron Larmor-precesses about the electrostatic field **E** that it produces in its motion: both the vectors **l** and **s** precess around their mechanical resultant **j** with the spin precessing twice as fast because of its anomalous (**d. Larmor's theorem** – violating) magnetic moment.]

(iii) Decrease in the number of states during the building-up process (riddle of statistical weights). [The half-integral unit of the apparent decrease is once again nothing but the contribution of the valence electron's half-integral spin to the total angular momentum of the atom. No violation of (**a. Bohr's building-up principle**).]

(iv) Higher multiplets [similar spin–orbit couplings are at work in the fine structure of higher multiplets].

(v) In the presence of an external weak field (anomalous Zeeman effect) each spectral line of a doublet or higher multiplet splits up into a number of

## 3.3 From the atomic core model to the electron's Zweideutigkeit 109

components, whose separation was empirically accounted for by (**e. Landé g factors**). [Again in semi-classical terms, the Zeeman patterns are due to **j** precessing around **H**, with both **l** and **s** in turn Larmor-precessing around **j**. The precessional angular velocity of the atom is given by $g$ (**e. Landé g factors**) times the Larmor orbital angular velocity (**d. Larmor's theorem**).] By (**c. Sommerfeld's space quantization**) the projection of the total angular momentum **j** on **H** can take only discrete values given by $mh/2\pi$.

(vi) In the presence of a strong magnetic field, new spectral patterns appear (Paschen–Back effect). [This is because the coupling between **l** and **s** breaks down, and **l** and **s** precess independently of each other around **j** (**d. Larmor's theorem**).]

∴ The electron runs about in two different (half-integral valued) states with the same (integral valued) orbital angular momentum $k$ (electron's *Zweideutigkeit* – electron's spin $\pm 1/2$ in unit of $h/2\pi$)

---

[a] In the formula: $T$ = energy level, $\alpha$ = fine structure constant, $R$ = Rydberg constant, $\sigma$ and $s$ = screening constants, $n$ = principal quantum number.
[b] The electron moves in the electrostatic field $E$ created by the protons of the nucleus. Then, by special relativity, in the electron frame there appears a magnetic field $B$ with which the electron's spin magnetic moment interacts. The spin–relativity doublets were called in the 1920s 'regular doublets' and the intervals between them in various elements of the periodic table were known accurately from the observed differences between a number of X-ray spectral lines.

---

Table 3.3, nested to the demonstrative induction of Table 3.1. Of course, the rule was not *demonstrated*, i.e. it was not a theorem deducible from quantum theoretical axioms, as Pauli himself repeatedly complained. Yet it was not a bold hypothesis plucked out of the air either. The new rule was well grounded in spectroscopic phenomena, in a series of phenomenological laws, as well as in the newly introduced concept of the electron's *Zweideutigkeit*. Without the previous introduction of the electron's *Zweideutigkeit*, the exclusion rule could not have followed. No wonder Pauli presented the new rule as an important consequence of his breakthrough.

Once again, the passage from the old to the new quantum theory need not be as drastic as the Kuhnian picture may suggest. New nomic generalizations may be introduced by deriving them from anomalous phenomena, old theoretical assumptions suitably reinterpreted as well as brand new scientific concepts. The prospective intelligibility of the electron's spin and Pauli's exclusion rule is grounded in the theoretical and experimental *milieu* of the old quantum theory. Anomalous phenomena and old theoretical assumptions evolved together, almost symbiotically, and together came to be modified and reshaped. In this way, new scientific concepts and nomic generalizations could emerge and become live options. Like Newton, Pauli too did not feign hypotheses.

Table 3.3

**Phenomenal premise**
(1) The experimentally found number of sublevels into which an $n_k$-spectral term splits in the presence of a strong magnetic field is $2(2k-1)$.

**Major premises**
(2) The largest number of sublevels into which an $n_k$-spectral term splits in the presence of a strong magnetic field is equal to the largest number of electrons per $n_k$-orbit (*Stoner's rule reinterpreted via Pauli's permanence of g sums*).
(3) The largest number of electrons per $n_k$-orbit is equal to $2n^2$ (*Rydberg's rule for the closure of electronic groups*).
(4) The number of states $2n^2$ (i.e. $2(2k-1)$) is due to the non-mechanical ability of the valence electron to run about in two different states with the same orbital angular momentum $k$ (electron's *Zweideutigkeit*).

∴ Any two electrons with the same orbital angular momentum state $k$ must be in two (*zweideutig*) different states for the electronic group to be closed (Pauli's exclusion rule).

Practitioners of the old quantum theory were in the position to appreciate and to opt for the electron's *Zweideutigkeit* and the new nomic generalization associated with it (Pauli's exclusion rule). This choice was a *rational* step, in the Enlightenment's sense that Ernst Cassirer (Section 1.4.2) contributed much to re-evaluate: namely, the new concept and nomic generalization brought to light the regularity immanent in prima facie anomalous phenomena, unveiled their inherent 'reason'. No wonder Newton's method of deduction from phenomena exerted an enormous influence on the *esprit systématique* of the *philosophes*.[47] What the Enlightenment strenuously advocated was not a trans-historical idea of scientific rationality, but rather the idea of a 'reason' immanent in phenomena and gradually unfolding across scientific developments. On the other side, the Encyclopaedists were also well aware that 'the general system of the sciences ... is a sort of labyrinth, a tortuous road which the intellect enters without quite knowing what direction to take'.[48] Scientific knowledge is

---
[47] Jean Le Rond d'Alembert, one of the leading figures of the French Enlightenment, was a professed disciple of Newton. In his *Preliminary Discourse to the Encyclopaedia of Diderot*, he proclaimed that 'the single true method of philosophising as physical scientists consists either in the application of mathematical analysis to experiments, or in observation alone, enlightened by the spirit of method, aided sometimes by conjectures when they can furnish some insights, but rigidly disassociated from any arbitrary hypotheses.' D'Alembert (1751). English translation (1995), pp. 24–5.
[48] D'Alembert (1751). English translation (1995), p. 46.

a labyrinth. But there is a hidden 'reason' in the apparent jungle of experimental data and anomalous phenomena. Scientific lexicons are only tentative and provisional maps to find our way out through the labyrinth.

# 4

# How Pauli's rule became the exclusion principle: from Fermi–Dirac statistics to the spin–statistics theorem

Shortly after their introduction, the electron's *Zweideutigkeit* and Pauli's *Ausschliessungsregel* were embedded into a growing theoretical framework: from Fermi–Dirac statistics in 1926 (Section 4.2), to the non-relativistic quantum mechanics of the magnetic electron with Pauli's spin matrices in 1927 (Section 4.3); from Wigner and von Neumann's group theoretical derivation of the spin matrices in 1927 (Section 4.4), to Jordan's reinterpretation of Pauli's exclusion principle in terms of anticommutation relations for particle creation and annihilation operators in 1928 (Section 4.5), which paved the way for quantum field theory. The most important step in building up this theoretical framework was the transition from non-relativistic to relativistic quantum mechanics, with Dirac's equation for the electron in 1928 (Section 4.6), which finally allowed the derivation of the electron's spin with its anomalous magnetic moment and hence clarified in a conclusive way the origin of spectroscopic anomalies. The negative energy solutions of the Dirac equation, and Dirac's attempt to accommodate them via the hole theory in 1930, anticipated the experimental discovery of the antiparticle of the electron, the positron. Sections 4.7–4.9 reconstruct Pauli's sustained criticism of Dirac's hole theory. The history of this debate is intertwined with Pauli's search for a spin–statistics connection, which finally culminated in the spin–statistics theorem in 1940. With this theorem, the nomological shift from the status of a phenomenological rule to that of a fundamental scientific principle is finally completed.

## 4.1 Introduction

The year 1925 was the *annus mirabilis* for quantum mechanics. In July 1925 Heisenberg[1] arrived at his famous *Umdeutung* (reinterpretation) of

---

[1] Heisenberg (1925b).

## 4.1 Introduction

classical mechanics: the new mechanics involved matrices, as Born soon recognized; hence the subsequent development of matrix mechanics in two pioneering papers, due to Born and Jordan, and to Born, Heisenberg, and Jordan.[2] Independently of Born, in November 1925, Dirac[3] laid down a new formalism implying the same commutation relations

$$i(pq - qp) = h \qquad (4.1)$$

where $p$ denotes momentum, and $q$ denotes position. The formalism was successfully applied to the hydrogen atom.[4] In January 1926, Pauli, who had just completed a monograph on the old quantum theory – his so-called Old Testament[5] – made a contribution to the theory of the hydrogen atom[6] from the point of view of the new formalism developed by Born, Heisenberg, and Jordan. Pauli briefly mentioned at the end of this paper Uhlenbeck and Goudsmit's hypothesis of the spinning electron, and the discrepancy between the observed splitting and the calculated fine structure, a discrepancy that Thomas solved one month later, in February 1926, as mentioned in Section 2.3. In the meantime, the Balmer formula for the hydrogen atom had also been derived by Erwin Schrödinger[7] following an alternative formalism, inspired by Louis de Broglie's idea of matter waves.[8] In March 1926, Schrödinger proved the equivalence between the Born–Heisenberg–Jordan matrix mechanics and his own wave mechanics.[9]

Thus, by the summer of 1926, the new quantum theory had been laid down. Non-relativistic quantum mechanics took the place of that tentative mixture of classical physics and quantum concepts, phenomenological laws and numerology typical of the old quantum theory. In 1928, with Dirac's relativistic equation for the electron, the way was open to relativistic quantum mechanics. This theoretical transition was accomplished thanks to the many contributions of Werner Heisenberg, Erwin Schrödinger, Paul Dirac, Max Born, Pascual Jordan, and Wolfgang Pauli, among others. A historical reconstruction of the rise of the new quantum mechanics goes far beyond the scope and purpose of this book, and I shall not attempt it here.[10] I focus instead on a single aspect of this complex and multifaceted process: namely, how Pauli's rule came to be

---

[2] Born and Jordan (1925); Born, Heisenberg, and Jordan (1926).   [3] Dirac (1925).
[4] Dirac (1926a).   [5] Pauli (1926a).   [6] Pauli (1926b).
[7] Schrödinger (1926a), (1926b).   [8] de Broglie (1925).   [9] Schrödinger (1926c).
[10] A detailed historical reconstruction can be found in Mehra and Rechenberg (1982a), (1982b), (1982c).

embedded in this theoretical framework and hence became an important scientific principle.

Out of the numerous low-level generalizations and tentative phenomenological laws that constituted the old quantum theory, many were discarded. Heisenberg's *sharing principle*, Landé and Heisenberg's *branching rule*, Bohr's rule for electronic distribution and *unmechanischer Zwang* were simply rules of thumb, useful to get things right but without any theoretical justification. Others were retained but only as a ladder that can be thrown away once one has climbed up it: Landé's formula for $g$ factors, Pauli's rule of the permanence of $g$ sums, Stoner's rule, all were introduced on an empirical basis in order to allow calculations of relevant parameters. So also was Pauli's *Ausschliessungsregel*, following from Stoner's rule reinterpreted in the light of the electron's *Zweideutigkeit*. How could such a tentative phenomenological rule gain the status of an important scientific principle?

The aim of this chapter is to shed light on the process through which Pauli's rule gained that status. This process consisted in the gradual embedding of the empirical phenomena that had beset practitioners of the old quantum theory into the new theoretical framework – wherein Pauli's rule, suitably reformulated, acquired the necessity and nomological strength of a scientific principle. Pascual Jordan was one of the first to appreciate the importance of Pauli's rule and to call it 'Pauli's principle' as early as 1925.[11] Yet the testimony of physicists is not necessarily a reliable indicator of this nomological transition, given the terminological confusion that persisted in the literature around 1925–8.[12] A look at the theoretical steps that brought about this nomological shift is more instructive.

As we have seen in Chapter 2, in 1925 Uhlenbeck and Goudsmit reinterpreted the electron's *Zweideutigkeit* in terms of the semi-classical spinning electron model. But it was only in 1928, with Dirac's relativistic equation for the electron, that the spin angular momentum and magnetic moment could finally be derived. Dirac's equation nevertheless raised a serious conundrum: it allowed negative energy solutions. To circumvent this problem, in 1930 Dirac advanced a daring proposal, the hole theory, in which the exclusion principle played a major role to guarantee that the total energy of the system was positive definite. The hole theory was the

---

[11] Jordan (1925), footnote 1, p. 568.
[12] Charles Enz (2002, p. 128) attributes to Dirac (1926b) the merit of having coined the expression 'Pauli's exclusion principle'. Fermi (1926) still referred to it as a 'hypothesis'; while one year later, Pauli (1927b) adopted the alternative expression '*Äquivalenzverbot*', used also by Jordan in his joint paper with Wigner (1928).

beginning of a long-lasting conceptual battle between Dirac and some of his critics, Pauli among them. I shall reconstruct the history of this debate in this chapter: from the Pauli–Weisskopf 'anti-Dirac' paper in 1934, to Pauli's first incomplete proof of the spin–statistics theorem in 1936, until the final proof in 1940. Although Pauli's proof of the spin–statistics theorem was animated by anti-Dirac intentions, these were not fully realized in Pauli's final proof, which was still anchored to the hole theory via the requirement of positive energy.

It is via this series of theoretical steps that the *Ausschliessungsregel* ceased to be an isolated phenomenological rule for the closure of electronic shells, and came to be embedded into a growing theoretical framework, from which ultimately the spin–statistics theorem was derived. This theorem widened the nomological scope of Pauli's principle, which turned out to apply not only to electrons, but to *any* spin-1/2 particle, including positrons, protons, neutrons, muons, among many other not yet discovered particles. Some far-reaching consequences of this enlarged nomological status will be analysed in the next chapter, as far as the development of quantum chromodynamics in the 1960s is concerned.

It took fifteen years (from 1925 to 1940) to establish the nomological status of the exclusion principle as we now know it. In that span, quantum electrodynamics and quantum field theory were also developed. Their elaboration is intertwined with the history of Pauli's principle. So also is the history of quantum statistics; the application of group theory to quantum mechanics; and the passage from non-relativistic to relativistic quantum mechanics that I am going to reconstruct in the following sections.

## 4.2 Pauli's rule prescribes a new exclusion: Fermi–Dirac statistics

The exclusion rule struck as heaven sent. Neither in the original letter to Landé nor in the article where it was publicly announced, did Pauli offer a proof of the rule. The lack of a proof beset Pauli for many years. Again in his Nobel Prize Lecture, Pauli remarked that

> already in my original paper I stressed the circumstance that I was unable to give a logical reason for the exclusion principle or to deduce it from more general assumptions. I had always the feeling and I still have it today, that this is a deficiency. Of course in the beginning I hoped that the new quantum mechanics, with the help of which it was possible to deduce so many half-empirical formal rules in use at that time, will also rigorously deduce the exclusion principle.

> Instead of it there was for electrons still an exclusion: not of particular states any longer, but of whole classes of states, namely the exclusion of all classes different from the antisymmetrical one.[13]

The new exclusion became clear with the development of quantum statistics. As early as 1924, first Bose,[14] then Einstein[15] shortly after, elaborated the statistics for a photon gas. In 1926 it was the turn of Fermi,[16] and a few months later of Dirac,[17] to formulate the quantum statistics of an ideal gas of particles obeying Pauli's exclusion rule. Starting from classical thermodynamics, Fermi calculated the specific heat and the energy distribution of an ideal gas whose molecular motion was quantized so that each molecule could be thought of as a harmonic oscillator. Fermi introduced as his only assumption Pauli's rule:

> We now intend to investigate whether a similar hypothesis could ever give good results in the problem of the quantization of an ideal gas: we will therefore assume that at most one molecule whose motion is characterised by a particular set of quantum numbers can be contained in our gas; and we will show that this hypothesis leads to a perfectly consequent theory of the quantization of an ideal gas.[18]

The quantum state of any molecule was associated with three quantum numbers $s_1$, $s_2$, $s_3$ and Pauli's rule applied to it now read as follows: there cannot be two molecules in the whole gas with the same quantum numbers $s_1$, $s_2$, $s_3$. The quantized energy of each molecule was given by

$$h\nu(s_1 + s_2 + s_3) = h\nu s \quad (4.2)$$

Let a molecule with the energy $h\nu s$ be an $s$-molecule. Fermi investigated how a fixed value of energy $W = h\nu E$ (where $E$ is a whole number) could be divided among a number $N_s$ of $s$-molecules, whereby the number $N_0$ has zero energy; $N_1$ has energy $h\nu$; $N_2$ has energy $2h\nu$, and so forth. The novelty with respect to the classical Maxwell–Boltzmann statistics is that an interchange of molecules did not result in two different orderings, but it counted as a single ordering. While Fermi essentially followed classical mechanics in his description of the ideal gas, seven months later Dirac derived the same statistical result within the new framework of quantum mechanics.

In a pioneering article, where among other things Schrödinger's wave mechanics and perturbation theory were also discussed, Dirac considered the application of Heisenberg's matrix mechanics to many-particle systems:

---

[13] Pauli (1948), p. 136.   [14] Bose (1924).   [15] Einstein (1924), (1925a), (1925b).
[16] Fermi (1926).   [17] Dirac (1926b).   [18] Fermi (1926). Reprinted in Fermi (1962), p. 183.

## 4.2 Fermi–Dirac statistics

In Heisenberg's matrix mechanics it is assumed that the elements of the matrices that represent the dynamical variables determine the frequencies and intensities of the components of radiation emitted. The theory thus enables one to calculate just those quantities that are of physical importance, and gives no information about quantities such as orbital frequencies that one can never hope to measure experimentally. We should expect this very satisfactory characteristic to persist in all future developments of the theory.[19]

Dirac went on to consider a system with two identical particles, namely an atom with two electrons, one in the orbit $m$ and the other in the orbit $n$ so that the state of the atom is given by ($mn$). He then asked whether the state that results from the interchange of the two electrons ($nm$), and which is physically indistinguishable from ($mn$), should be counted as a second different state or as one and the same physical state as ($mn$). In Heisenberg's formalism, this would result in either a $2 \times 2$ matrix or in a matrix with only one row and one column, respectively. If they were two different states, one would be able to calculate the intensities due to the transitions ($mn$) → ($m'n'$) and ($nm$) → ($n'm'$) separately. But this is not the case since the two transitions are physically indistinguishable, and only their sum can be experimentally determined. Hence the conclusion:

> In order to keep the essential characteristic of the theory that it shall enable one to calculate only observable quantities, one must adopt the second alternative that ($mn$) and ($nm$) count as only one state.[20]

Thus, in the name of the aforementioned characteristic of Heisenberg's formalism, which confined attention to experimentally measurable quantities only, the states ($mn$) and ($nm$) counted as one. This had far-reaching consequences. By neglecting the interactions between the electrons and representing the state $m$ of electron 1 with the eigenfunction $\psi_m(x_1, y_1, z_1, t)$, and similarly for electron 2 in state $n$, the state ($mn$) of the atom could be described in terms of the product

$$\psi_m(x_1, y_1, z_1, t)\psi_n(x_2, y_2, z_2, t) = \psi_m(1)\psi_n(2) \qquad (4.3)$$

But the same state of the atom can also be described by the product $\psi_m(2)\psi_n(1)$, if the states ($mn$) and ($nm$) count as one. Thus, if there is to be a matrix with only one row and one column, it becomes necessary to find a set of eigenfunctions $\psi_{mn}$ of the form

$$\psi_{mn} = a_{mn}\psi_m(1)\psi_n(2) + b_{mn}\psi_m(2)\psi_n(1) \qquad (4.4)$$

---

[19] Dirac (1926b), p. 666–7.  [20] Ibid., p. 667.

that contains only one $\psi_{mn}$ describing both (mn) and (nm), and where the $a_{mn}$ and $b_{mn}$ are constants. As Dirac noticed, there are only two ways of choosing this set so as to satisfy this requirement: either (1) $a_{mn} = b_{mn}$ so that $\psi_{mn}$ is a symmetric function; or (2) $a_{mn} = -b_{mn}$ so that $\psi_{mn}$ is an antisymmetric function. But, as Dirac observed 'the theory at present is incapable of deciding which solution is the correct one.'[21] Dirac then associated antisymmetric functions with Pauli's exclusion rule:

> An antisymmetrical eigenfunction vanishes identically when two of the electrons are in the same orbit. This means that in the solution of the problem with antisymmetrical eigenfunctions there can be no stationary states with two or more electrons in the same orbit, which is just Pauli's exclusion principle. The solution with symmetrical eigenfunctions, on the other hand, allows any number of electrons to be in the same orbit, so that this solution cannot be the correct one for the problem of electrons in an atom.[22]

In a footnote at the end of this passage Dirac acknowledged the independent work of Heisenberg, who in May 1926 had arrived at the same result about Pauli's rule while working on the singlet–triplet splitting and the problem of ortho- and parhelium.[23] For the first time, Pauli's rule was pulled out of its original spectroscopic context and applied to quantum statistics for many-particle systems. Pauli's *Verbot* was reformulated in terms of antisymmetric functions for a system of two or more electrons.

Dirac went on to apply this result to an ideal gas of molecules, and reproduced Fermi's results. For a system of non-interacting gas molecules, one can take the product of the eigenfunctions of the single molecules. An assumption of equiprobability was introduced that resulted in different probability values, depending on whether the assembly of molecules was symmetrized or antisymmetrized: in the first case the Bose–Einstein statistics followed, in the second the Fermi–Dirac statistics. Thus, antisymmetric functions encoding Pauli's *Verbot* resulted in a new quantum statistics, which diverged from the classical Maxwell–Boltzmann statistics as well as from the Bose–Einstein quantum statistics. Dirac remarked that 'the solution with antisymmetrical eigenfunctions ... is probably the correct one for gas molecules, since it is known to be the correct one for electrons in an atom, and one would expect molecules to resemble electrons more closely than light-quanta'.[24] There was still a long way to go for the discovery of a connection between the nature of particles on the one side, and the quantum statistics they accordingly follow, on the other side. A quantum mechanical

---

[21] *Ibid.*, p. 669.   [22] *Ibid.*, p. 669–70.   [23] Heisenberg (1926).   [24] *Ibid.*, p. 672.

## 4.3 The non-relativistic quantum mechanics of the magnetic electron: Pauli's spin matrices

reinterpretation of the semi-classical spinning electron model was first required. And this crucial step was taken once again by Pauli.

By March 1926 Pauli had completely abandoned his original reluctance towards the spinning electron model, after Thomas's result on the anomalous factor 2 (Section 2.3). In a letter to Wentzel on 5 December 1926,[25] he announced a new paper[26] that extended Fermi–Dirac statistics to atoms with angular momentum and placed in an external magnetic field.

In that paper, Pauli referred back to a recent work of Heisenberg on the helium atom.[27] Heisenberg had successfully applied Pauli's exclusion rule to the helium atom, whose spectral pattern (singlet and triplet) had beset physicists for several years. He introduced, for a system of two electrons without spin, both symmetric and antisymmetric eigenfunctions, corresponding to parhelium and orthohelium, respectively. Every state of the system could then give rise to four states, three of which were symmetric in the spin coordinates and one antisymmetric. The observed singlet and triplet were traced back, respectively, to the parhelium symmetric eigenfunction being multiplied by the antisymmetric eigenfunction of the spin coordinates, and to the orthohelium antisymmetric eigenfunction being multiplied by the three symmetric eigenfunctions of the spin coordinates. Heisenberg's work preceded by one year Pauli's result on the spin matrices, and hence no precise formulation could yet be given of the 'spin coordinates'. No wonder Heisenberg presented his work in a letter to Pauli 'with doubtful feelings and no satisfaction. The calculations are all so imprecise and incomplete, the tidiest one is the fine structure, which comes out right'.[28]

Pauli commented on Heisenberg's result with the following words: 'it is then proved that quantum mechanics does not contradict the equivalence rule',[29] where 'equivalence rule' [*Äquivalenzregel*][30] is the new name that Pauli gave to his exclusion rule. In his paper, Pauli went on to repeat

---

[25] In Pauli (1979), p. 361.   [26] Pauli (1927a).   [27] Heisenberg (1926).
[28] Heisenberg to Pauli, 28 July 1926. In Pauli (1979), p. 338.   [29] Pauli (1927a), p. 83.
[30] 'According to the physical interpretation of the four quantum numbers due to Goudsmit and Uhlenbeck as a self-magnetism of the electron, we can express the sense of this rule as follows: there cannot be two electrons in a physically realised quantum state of an atom that are kinematically completely equivalent both in their motion and in their self-magnetism. Henceforth, this rule is designated briefly as the "equivalence rule".' *Ibid.*, p. 82.

Fermi's calculations, and investigated the magnetic behaviour of a degenerate gas, whose atoms followed the 'equivalence rule'. The results obtained for temperature-independent paramagnetism were in agreement with the empirical findings for alkali metals. Yet Pauli noticed that 'we still lack a theoretically satisfactory foundation for this special selection from the solutions of the quantum mechanical equations and also for the equivalence rule'.[31] While a theoretical foundation for Pauli's rule continued to prove elusive,[32] in the meantime important steps were taken towards a quantum mechanical reinterpretation of the spinning electron.

In March 1926, Heisenberg and Jordan[33] offered an analysis of the doublet fine structure and of the anomalous Zeeman effect in the light of the new matrix mechanics. The spin angular momentum was associated with a vector **s** whose components $s_x, s_y, s_z$ satisfied commutation relations in analogy with those for the orbital angular momentum vector **k**. The spin–orbit coupling appeared as a correction term proportional to $\mathbf{k} \cdot \mathbf{s}$ in the Hamiltonian, while the Larmor-violating magnetic interaction energy was given by

$$(e/2m_0c)H \cdot (\mathbf{k} + 2\mathbf{s}) \tag{4.5}$$

where the anomalous factor 2, which had beset Landé and was originally ascribed to a core magnetic anomaly, was here ascribed to an anomalous spin magnetic moment. As we shall see in Section 4.6, this factor could finally be derived from Dirac's relativistic equation.

In February 1927, it was the turn of Charles Galton Darwin[34] to reinterpret the spinning electron model as a vector wave in the light of wave mechanics. There were however a few difficulties: the wave equation for the electron spin implied a six-dimensional space that escaped any possibility of visualization; furthermore, since the stationary states corresponded to angular momenta that were integral multiples of $h/2\pi$, it was impossible to get the spin angular momentum value $\pm 1/2$ in units of $h/2\pi$. To accommodate the latter difficulty, Darwin assumed that the wave

---

[31] *Ibid.*, p. 83.
[32] In January 1927, Ehrenfest took up the challenge and tried to provide Pauli's rule with a theoretical foundation. Ehrenfest's (1927) idea was to found the exclusion rule on the impenetrability of matter: he believed that electrons with the same spin were penetrable to each other because of the strong magnetic attraction that supposedly overcame the electrostatic repulsion, while opposite spins implied magnetic repulsion and hence impenetrability. But the idea was based on a theoretical mistake, as Ehrenfest himself realized, and the paper was soon withdrawn. See Ehrenfest's letter to Pauli, 24 January 1927, in Pauli (1979), p. 371.
[33] Heisenberg and Jordan (1926). [34] Darwin (1927a).

function associated with each state was not single-valued, but double-valued. By introducing double-valued functions defining a unit vector, he obtained the formulas for the intensities of spectral lines and for the magnetic moment of the atom, and could also extend his theory to many-electron atoms.[35] These mathematical results agreed with those that Pauli had independently obtained in another pioneering article.

In contrast with Darwin, Pauli[36] did not resort to vector waves. He introduced instead a new independent variable for the spin component $s_z$, beside the space coordinates $q_k$, in the wave function $\psi$. The variable $s_z$ could take only the two characteristic values $+1/2$, and $-1/2$ in units of $h/2\pi$. Thus, the wave function $\psi(q_k, s_z)$ has two components $\psi_\alpha(q_k)$ and $\psi_\beta(q_k)$ corresponding to the spin values $+1/2(h/2\pi)$, and $-1/2(h/2\pi)$. The squares of the absolute values

$$|\psi_\alpha(q_k)|^2 \cdot dq_1\, dq_2\, dq_3 \quad \text{and} \quad |\psi_\beta(q_k)|^2 \cdot dq_1\, dq_2\, dq_3 \qquad (4.6)$$

give the probability density that in the quantum state where the space coordinate $q_k$ lies in the infinitesimal interval $(q_k, q_k + dq_k)$, the spin is respectively either up or down. To incorporate the spin into Schrödinger's equation, the magnetic interaction energy due to the spin magnetic moment interacting with the internal magnetic field created by the electrostatic field of the nucleus, had to be introduced. To this purpose, along the lines of Heisenberg and Jordan,[37] Pauli introduced the spin angular momentum components $s_x, s_y, s_z$ where he set

$$s_x = \frac{1}{2}\sigma_x \qquad s_y = \frac{1}{2}\sigma_y \qquad s_z = \frac{1}{2}\sigma_z \qquad (4.7)$$

and $\sigma$ are the $2 \times 2$ matrices

$$\sigma_x = \begin{pmatrix} 0 & 1 \\ 1 & 0 \end{pmatrix} \quad \sigma_y = \begin{pmatrix} 0 & -i \\ i & 0 \end{pmatrix} \quad \sigma_z = \begin{pmatrix} 1 & 0 \\ 0 & -1 \end{pmatrix} \qquad (4.8)$$

which have since become known as the Pauli spin matrices. They satisfy the relations

$$\begin{aligned} \sigma_\mu^2 &= 1 & (\mu = x, y, z) \\ \sigma_\mu \sigma_\nu + \sigma_\nu \sigma_\mu &= 0 & (\mu \neq \nu; \mu, \nu = x, y, z) \end{aligned} \qquad (4.9)$$

Equipped with these spin matrices, Pauli went on to calculate the spin–orbit correction term for the Hamiltonian. The mathematical results thus

---

[35] Darwin (1927b).  [36] Pauli (1927b).  [37] Heisenberg and Jordan (1926).

obtained for the 'magnetic electron' were identical with Darwin's. Pauli extended them to an assembly of electrons satisfying the 'equivalence rule'. Despite the important results, the theory still introduced the spin half-integral value and Thomas's relativistic correction 'by hand', so to speak. Hence, Pauli concluded the paper with the apologetic remark that the theory was only provisional because no Lorentz invariant formulation had yet been found. Nor did Darwin's alternative theory fare any better on this score.[38] Nevertheless, as van der Waerden has remarked:

> It seems that Pauli underestimated the importance and the final character of his methods and results ... Pauli's matrices were used by Dirac to form a relativistic first-order wave equation ... The step from one to two $\psi$ components is large, whereas the step from two to four components is small ... it was Pauli who made the first decisive step, and in this part of his paper there is nothing provisional or approximate.[39]

## 4.4 Group theory enters the scene

The next important step in building up the new framework was the application of group theory to quantum mechanics in 1927–8. The upshot consisted in the derivation of Pauli's spin matrices from the invariance with respect to space rotation. The University of Göttingen was the main centre for group theory. Here, Max Born had developed matrix mechanics (the *Göttingen Mechanik* as Pauli called it); David Hilbert and his school had been active; and in 1918 Emmy Noether had formulated the famous theorem establishing a connection between symmetries and conservation laws. And it was in Göttingen that between the late 1920s and the beginning of the 1930s Hermann Weyl and Eugene Wigner worked out an analysis of the recent quantum mechanical results from the point of view of group theory. In 1928 Weyl published his fundamental volume[40] on the subject, which had been preceded in 1927 by a pioneering work of Wigner.

---

[38] 'Apart from the greater elegance of the formulae, the real advantage of the vector conception is that it should more readily suggest extensions of application. This it undoubtedly will do, but in the most important such extension, the relativity transformation, the simplest form of the generalization gives a wrong result ... From this point of view all that we can at present claim is that the vector principle helps to make explicit the widespread difficulties in the way of uniting the quantum theory with relativity.' Darwin (1927b), p. 245.

[39] Van der Waerden (1960), p. 223.   [40] Weyl (1928).

Following on Heisenberg's work on the helium atom,[41] Wigner[42] extended Pauli's exclusion rule to the case of an ensemble of three, and more generally $n$, electrons without spin. The calculation turned out to be so laborious and complex that Wigner asked the mathematician von Neumann for advice. Von Neumann referred him back to the work of Frobenius and Schur on group theory. Despite the breakthrough of applying group theory to quantum mechanics, Wigner's 1927 work was incomplete, as it did not take into account the spin. A full-blown application of group theory to quantum mechanics was carried out in the von Neumann and Wigner trilogy,[43] between December 1927 and June 1928. They derived the electron kinematics and various spectroscopic phenomena (from the selection rules and intensity rules for spectral terms to the Landé $g$ factors for the Zeeman effect) by applying group theory to Pauli's analysis of the magnetic electron.[44] In particular, they could retrieve Pauli's spin matrices from space rotation.

Wigner and von Neumann's contribution opened up a fruitful avenue of research showing how quantum mechanical properties could be elegantly derived from group theory. But this new strand did not meet with universal acceptance. There was a widespread reluctance among physicists to engage in the highly abstract and unfamiliar mathematics of group theory, teasingly dubbed the 'group pest' [*Gruppenpest*].[45] For instance, in November 1929 J. C. Slater published a thorough compendium to the theory of complex spectra where he emphasised that his calculations did not resort to group theory, and 'in contrast, no mathematics but the simplest is required'.[46] Another crucial application of group theory to quantum mechanics, namely the Lorentz invariance of the electron wave function, was just around the corner. Its discovery opened the era of relativistic quantum mechanics. Before turning to it, let us first have a look at another important step in the build-up of the new framework.

## 4.5  From quantum electrodynamics to quantum field theory: the exclusion principle re-expressed in terms of anticommutation relations

On 2 February 1927, Dirac submitted a paper[47] that was destined to exert an enormous influence on the theoretical developments of the following

---
[41] Heisenberg (1926).   [42] Wigner (1927).
[43] Von Neumann and Wigner (1928a), (1928b), (1928c).   [44] Pauli (1927b).
[45] See, for instance, Ehrenfest's letter to Pauli, 22 September 1928. In Pauli (1979), p. 474.
[46] Slater (1929), p. 1294.   [47] Dirac (1927).

decade. The paper began with an acknowledgment of the recent progress of new quantum theory as far as dynamics was concerned, to which however no similar progress corresponded on the side of electrodynamics. Dirac's paper aimed to remedy this deficiency, and, indeed, it paved the way to quantum electrodynamics.

In classical electrodynamics, in the case of no interaction between the field and the atom, the Hamiltonian of the system is given by

$$H = \sum_r E_r + H_0 \quad (4.10)$$

where $H_0$ is the Hamiltonian for the atom alone, and the radiation field is resolved into its Fourier components $r$ having energies $E_r$ and phases $\theta_r$. The interaction between the atom and the field is expressed by adding an interaction term to the above equation. Dirac gave an analogous treatment for quantum electrodynamics, where the variables $E_r$ and $\theta_r$ become $q$-numbers satisfying the standard quantum condition

$$\theta_r E_r - E_r \theta_r = ih \quad (4.11)$$

Thus, the radiation field acquires light-quantum properties: the energy of its components is quantized and takes only integral multiples of the quantum of action $h\nu_r$. A far-reaching consequence of this move was that the Hamiltonian describing the interaction between the atom and the field turned out to be identical to the Hamiltonian describing the interaction between an atom and an assembly of light quanta satisfying Bose–Einstein statistics (under a suitable choice of interaction energy).

Dirac defined the Hamiltonian for an assembly of systems, independent of each other and all perturbed in the same way, as the sum of the Hamiltonian for the unperturbed system $H_0$ plus the perturbing energy $V$. This satisfied the Schrödinger equation

$$ih \frac{\partial \psi}{\partial t} = (H_0 + V)\psi \quad (4.12)$$

whose solution is given by expanding $\psi$ into the eigenfunctions $\psi_r$ of the unperturbed system, i.e. $\psi = \Sigma_r a_r \psi_r$, each associated with one stationary state labelled by the suffix $r$, where the $a_r$ are functions of time only, and $|a_r|^2$ is the probability of the system being in state $r$ at any time.[48] The variables $a_r$ and $iha_r{}^*$ are canonical conjugates, and Dirac set them equal to

$$b_r = a_r \, e^{-iW_r t/h} \qquad b_r^* = a_r^* \, e^{iW_r t/h} \quad (4.13)$$

---

[48] Ibid., p. 248.

where $W_r$ is the energy of state $r$. The crucial step consisted in treating the new variables as $q$-numbers satisfying

$$\begin{aligned} b_r b_r^* - b_r^* b_r &= 1 \\ b_r b_s - b_s b_r &= 0 \\ b_r^* b_s^* - b_s^* b_r^* &= 0 \\ b_r b_s^* - b_s^* b_r &= 0 \end{aligned} \tag{4.14}$$

The interaction Hamiltonian for the perturbed system $V$ can now be expressed as $H_{rs} = W_r \delta_{rs} + v_{rs}$, which is a matrix element of the total Hamiltonian $H = H_0 + V$. The new variables were defined in terms of $N_r$ and $\theta_r$, themselves $q$-numbers and canonical conjugates, denoting respectively the number of systems in the state $r$, and the phase of the system:

$$b_r = N_r^{\frac{1}{2}} e^{-i\theta_r/h} \qquad b_r^* = N_r^{\frac{1}{2}} e^{i\theta_r/h} \tag{4.15}$$

so that the total Hamiltonian could be finally written as

$$F = \sum_r W_r N_r + \sum_{rs} v_{rs} N_r^{\frac{1}{2}} N_s^{\frac{1}{2}} e^{i(\theta_r - \theta_s)/h} \tag{4.16}$$

with the first term denoting the unperturbed system and the second the additional energy due to perturbation. Dirac then applied the result to systems obeying Bose–Einstein statistics (namely, photons). He set $b(r_1, r_2, \ldots)$ to be a symmetric function of the variables $r_1, r_2, \ldots$ so as to obtain Einstein's laws for the emission and absorption of radiation.

Some of the essential tools of the quantum field theory formalism were introduced in this pioneering paper: the new canonical variables (4.13) with their commutation relations (4.14) are nothing but the particle creation and annihilation operators; the variable $N_r$ is the occupation number operator that in the Fock representation of quantum field theory indicates the number of indistinguishable particles or quanta occupying the state $r$; and the overall technique, which Pauli teasingly called 'Dirac's acrobatics', is nothing but the second quantization that ever since then has been the distinctive technique of quantum field theory.[49] In this way, Dirac showed that a system of $N$ particles with symmetric wave functions can be treated as a matter field, upon second quantization, much as the radiation field shows light quantum properties: waves are particles no more than particles are waves. This old adage finally found a rigorous mathematical

---

[49] For details about Dirac's technique of second quantization, see Tomonaga (1997), pp. 97–107.

expression. Dirac's pioneering analysis was confined to particles obeying Bose–Einstein statistics, whose operators satisfied the commutation relations (4.14). The extension of this analysis to electrons, and in general to particles obeying Fermi–Dirac statistics proved more complicated. It was Pascual Jordan who took the first step in this direction.

Jordan had been a Ph.D. student of Max Born at Göttingen, actively involved in the elaboration of matrix mechanics in 1925–6. Between July 1927 and January 1928, he made a fundamental contribution to the development of quantum field theory. He introduced a brand new mathematical device. Instead of using commutation relations like those Dirac had used for the bosonic creation and annihilation operators, Jordan introduced a new type of algebraic relation for electrons and more generally particles obeying Fermi–Dirac statistics: the anticommutation relations. Jordan[50] followed Dirac and developed a corresponding theory for an ideal Fermi–Dirac gas. The annihilation and creation operators $b_r$ and $b_r^\dagger$ were defined in terms of the two conjugate quantities $\Theta_r$ and $N_r$

$$b_r = (1 - N_r)^{\frac{1}{2}} e^{-i\Theta_r}$$
$$b_r^\dagger = e^{i\Theta_r} (1 - N_r)^{\frac{1}{2}} \qquad (4.17)$$

where the occupation number operator $N_r$ could take as eigenvalues only 0 or 1, as required by Pauli's exclusion principle, by contrast with the Bose–Einstein case discussed by Dirac where $N_r$ could take 0, 1, 2, ... eigenvalues.

But the main novelty was introduced in a subsequent joint paper with Wigner, in January 1928, entitled 'On Pauli's *Äquivalenzverbot* [equivalence prohibition].'[51] An ideal gas obeying Fermi–Dirac statistics was described by means of a three-dimensional wave field, whose non-commutative properties of the wave amplitudes expressed Pauli's *Verbot*. The occupation number operator $N_r$ was defined as

$$N_r = b_r^\dagger b_r \qquad 1 - N_r = b_r b_r^\dagger \qquad (4.18)$$

with $\left(b_r^\dagger\right)^2 = (b_r)^2 = 0$, which expressed Pauli's prohibition for more than one fermion to be created or annihilated in the state $r$. Given then an assembly of fermions, let us denote two different states any fermion can be in as $q'$ and $q''$. The effect of the creation operator $b^\dagger(q')$ on the antisymmetrized state of the assembly depends on the number of occupied states $q''$ that come before $q'$ in the Heisenberg–Dirac determinantal wave

---

[50] Jordan (1927).  [51] Jordan and Wigner (1928).

function for the fermions, given an established order of the $q'$. Let us denote the product of all these quantities $\{1 - 2N(q'')\}$, for all $q'' \leq q'$, as

$$\nu(q') = \prod \{1 - 2N(q'')\} \quad (4.19)$$

Two new quantities were then introduced

$$\begin{aligned} a(q') &= \nu(q') \cdot b(q') \\ a^\dagger(q') &= b^\dagger(q') \cdot \nu(q') \end{aligned} \quad (4.20)$$

whose product was antisymmetric in $q'$ and $q''$

$$\begin{aligned} a(q')a(q'') &= -a(q'')a(q') \\ a^\dagger(q')a^\dagger(q'') &= -a^\dagger(q'')a^\dagger(q') \end{aligned} \quad (4.21)$$

and, most importantly, they satisfied the following new anticommutation relation

$$a^\dagger(q')a(q'') + a(q'')a^\dagger(q') = \delta(q' - q'') \quad (4.22)$$

For the first time, anticommutation relations were introduced, which ever since then have played a key role in the second quantization of fields obeying the Pauli principle. While bosonic fields are quantized according to the usual commutation relations $[b, b^\dagger]_- = bb^\dagger - b^\dagger b$ for the annihilation and creation operators, fermionic fields are quantized using the anticommutation relations $[a, a^\dagger]_+ = aa^\dagger + a^\dagger a$ originally introduced by Jordan and Wigner. In this way, quantum statistics came to be embedded in quantum field theory through the commutative/anticommutative multiplication properties of the particle and antiparticle creation–annihilation operators. Jordan and Wigner concluded their article with a comparison between the Bose–Einstein case and the Fermi–Dirac one:

> We have derived these equations in order from the outset to lay down the foundation of Pauli's *Äquivalenzverbot* ... We can say that the existence of material particles and the validity of the Pauli principle can be understood as a consequence of the quantum mechanical multiplication properties of the de Broglie wave amplitudes.[52]

The paper finished with the remark that this result could be easily generalized in a Lorentz invariant form, in analogy with the Lorentz invariant commutative relations for the charge-free electromagnetic field introduced by Jordan

---

[52] *Ibid.*, p. 640.

and Pauli in a recent work.[53] In fact it was only with Dirac's equation for the electron that relativistic quantum mechanics finally became a live option.

## 4.6 Towards relativistic quantum mechanics: the Dirac equation for the electron and the hole theory

In a letter dated 11 January 1928, Darwin informed Pauli that 'Dirac has got a new system of wave equations which does the whole spinning electron correctly, Thomas correction, relativity and all.'[54] Indeed, on 2 January, Dirac had submitted a paper where a Lorentz invariant equation for the electron was finally presented.[55] In the introductory paragraph, which is worth quoting in some detail, Dirac summarized the previous theoretical achievements and latched his contribution onto them:

> The new quantum mechanics, when applied to the problem of the structure of the atom with point-charge electrons, does not give results in agreement with experiment. The discrepancies consist of 'duplexity' phenomena, the observed number of stationary states for an electron in an atom being twice the number given by the theory. To meet the difficulty, Goudsmit and Uhlenbeck have introduced the idea of an electron with a spin angular momentum of half a quantum and a magnetic moment of one Bohr magneton. This model for the electron has been fitted into the new mechanics by Pauli (1927[b]), and Darwin (1927[b]), working with an equivalent theory, has shown that it gives results in agreement with experiment for hydrogen-like spectra to the first order of accuracy.
>
> The question remains as to why Nature should have chosen this particular model for the electron instead of being satisfied with the point-charge. One would like to find some incompleteness in the previous methods of applying quantum mechanics to the point-charge electron such that, when removed, the whole of the duplexity phenomena follow without arbitrary assumptions. In the present paper it is shown that this is the case, the incompleteness of the previous theories lying in their disagreement with relativity, or, alternatively, with the general transformation theory of quantum mechanics. It appears that the simplest Hamiltonian for a point-charge electron satisfying the requirements of both relativity and the general transformation theory leads to an explanation of all duplexity phenomena without further assumption. All the same there is a great deal of truth in the spinning electron model, at least as a first approximation.[56]

---

[53] Jordan and Pauli (1928).    [54] Darwin to Pauli, 11 January 1928. In Pauli (1979), p. 424.
[55] Dirac (1928a).    [56] *Ibid.*, p. 610.

## 4.6 Towards relativistic quantum mechanics

Dirac derived his equation for the electron in an attempt to overcome two difficulties affecting another relativistic wave equation

$$\left[\left(i h \frac{\partial}{c \partial t}+\frac{e}{c} A_{0}\right)^{2}+\sum_{r}\left(-i h \frac{\partial}{\partial x_{r}}+\frac{e}{c} A_{r}\right)^{2}+m^{2} c^{2}\right] \psi=0 \quad (4.23)$$

that Gordon[57] and independently Klein[58] had earlier introduced. The Klein–Gordon equation was a differential equation, second-order with respect to time, whose conjugate imaginary

> is the same as one would get if one put $-e$ for $e$ [in the original equation]. The wave equation thus refers equally well to an electron with charge $e$ as to one with charge $-e$ ... One gets over the difficulty in the classical theory by arbitrarily excluding those solutions that have a negative [energy] $W$. One cannot do this on the quantum theory, since in general a perturbation will cause transitions from states with $W$ positive to states with $W$ negative. Such a transition would appear experimentally as the electron suddenly changing its charge from $-e$ to $e$, a phenomenon which has not been observed. The true relativity wave equation should thus be such that its solutions split up into two non-combining sets, referring respectively to the charge $-e$ and $e$.[59]

Searching for a first-order differential equation, Dirac found the famous equation, which, in current notation, reads as

$$\left[i \sum_{\mu} \gamma^{\mu} p_{\mu}+m c\right] \psi=0 \quad (4.24)$$

where $\gamma^{\mu}$ are the four Dirac matrices with $\mu = 0, 1, 2, 3$ satisfying $\gamma^{\mu}\gamma^{\nu} + \gamma^{\nu}\gamma^{\mu} = 2g^{\mu\nu}$, where $g^{\mu\nu}$ is the contravariant Lorentz metric, and $\psi$ is Dirac spinor. Without entering into technical details, it suffices here to say that Dirac first introduced new matrices $\boldsymbol{\alpha}$ to be plugged into the sought-after first-order differential equation, and satisfying

$$\begin{aligned} \boldsymbol{\alpha}_{\mu}\boldsymbol{\alpha}_{\nu}+\boldsymbol{\alpha}_{\nu}\boldsymbol{\alpha}_{\mu}=0 \quad &\text{with } \mu \neq \nu \quad \text{and} \quad \nu, \mu=0, 1, 2, 3 \\ \boldsymbol{\alpha}_{\mu}^{2}=1& \end{aligned} \quad (4.25)$$

In order to find four $4 \times 4$ $\boldsymbol{\alpha}$ matrices satisfying condition (4.25), Dirac resorted to Pauli's three $2 \times 2$ spin matrices (4.8) and extended them in a direct product manner to bring in two more rows and columns. Having

---

[57] Gordon (1926).   [58] Klein (1927).   [59] Dirac (1928a), p. 612.

proved that the resultant wave equation was invariant under a Lorentz transformation, Dirac went on to show that the wave equation

> appears to be sufficient to account for all the duplexity phenomena. On account of the matrices ... containing four rows and columns, it will have four times as many solutions as the non-relativity wave equation, and twice as many as the previous relativity wave equation [Klein–Gordon]. Since half the solutions must be rejected as referring to the charge $+e$ on the electron, the correct number will be left to account for duplexity phenomena.[60]

Thus, the anomalous doubling of the states – encoded by the Landé $g$ factor 2, originally ascribed to a core magnetic anomaly and later to the electron's *Zweideutigkeit* – could be conclusively explained: the electron spin angular momentum and magnetic moment were no longer assumed, but they naturally followed from Dirac's relativistic wave equation. The Thomas relativistic correction factor too, which in the spinning electron model had been plugged in 'by hand', came up straight from the relativistic wave equation. Dirac could derive the energy levels for the electron motion in an external field, in agreement, at least as a first approximation, with the results already obtained by Pauli[61] and Darwin.[62]

In a subsequent paper, Dirac[63] derived the selection rules for the electron orbital angular momentum in agreement with the well-known empirical results for the alkali doublets. The relative intensities of the multiplet lines, the Zeeman effect and the Paschen–Back effect were also derived from Dirac's theory in a straightforward way. Later, Sommerfeld's formula for the fine structure of the hydrogen spectrum,[64] and the Klein–Nishina formula for the scattering of light by electrons[65] were similarly derived from it. All the anomalous spectroscopic phenomena that around 1921–4 marked the crisis of the old quantum theory could be finally explained within the relativistic framework of Dirac's theory of the electron.

Despite the empirical successes, there was however a major problem. The 'true' sought-after relativistic wave equation was expected to separate the negative energy solutions from the positive energy ones, by contrast with the Klein–Gordon equation, which gave a superposition of them. Unfortunately, the Dirac equation fared no better on this score. As Klein quickly retorted against Dirac, in the case of time-dependent external fields, transitions from positive to negative energy states also affected the Dirac equation (the Klein Paradox). Shortly after the breakthrough of 1928, Dirac returned to this problem.[66] Remarking that no

---

[60] *Ibid.*, p. 618.   [61] Pauli (1927b).   [62] Darwin (1927b).   [63] Dirac (1928b).
[64] See Darwin (1928), Gordon (1928).   [65] Klein and Nishina (1929).   [66] Dirac (1930a).

## 4.6 Towards relativistic quantum mechanics

hard-and-fast distinction between positive and negative energy solutions was available in quantum theory, Dirac identified the wave packet constituted by the superposition of the negative energy solutions as describing the motion of 'an electron of charge $+e$ (and positive energy) moving in the original electromagnetic field. Thus *an electron with negative energy moves in an external field as though it carries a positive charge.*'[67] Dirac was at this point very close to the introduction of the antiparticle of the electron: the negative energy solutions of his equation were associated with positive-charged electrons. Antiparticles would have been easily recognized, if Dirac had not followed Weyl's[68] mistaken identification of the negative energy electrons with protons. This identification was incompatible with the different masses of electrons and protons. As Dirac later recalled,[69] at that time he was more concerned with getting a satisfactory theory of the electron than with bringing in protons. The real upshot of the paper was indeed to bring in the exclusion principle via the so-called 'negative energy sea':

> The most stable states for an electron (i.e. the states of lowest energy) are those with negative energy and very high velocity. *All the electrons in the world will tend to fall into these states with emission of radiation. The Pauli exclusion principle, however, will come into play and prevent more than one electron going into any one state.* Let us assume there are so many electrons in the world that all the most stable states are occupied, or, more accurately, that *all the states of negative energy are occupied except perhaps a few of small velocity* ... We shall have an infinite number of electrons in negative energy states, and indeed an infinite number per unit volume all over the world, but if their distribution is exactly uniform we should expect them to be completely unobservable. *Only the small departures from exact uniformity, brought about by some of the negative energy states being unoccupied, can we hope to observe.*[70]

The few vacant states or 'holes' – as Dirac called them – in the negative energy sea were introduced by analogy with X-rays, emitted when an internal vacancy in the electronic configuration occurs. However, while in the X-ray context the holes count as objects of negative energy because ordinary positive energy electrons are required to fill them up, in the negative energy sea the holes counted as objects of positive energy because negative energy electrons were required to fill them up. The holes were supposed to behave like ordinary particles with positive charge $+e$: whenever an electron jumped from a negative energy state to a positive energy

---

[67] *Ibid.*, p. 361. Emphasis in the original.  [68] Weyl (1929).
[69] Kuhn's interview of Dirac (14 May 1963) in AHQP.
[70] Dirac (1930a), p. 362. Emphasis in the original.

one, it left behind a hole (electron/hole creation); vice versa, when it jumped from a positive energy to a negative energy state, there was a process of electron/hole annihilation.

The hole theory was a heuristically powerful tool to deal with the problem of negative energy states. But it faced several difficulties. First, the infinite number of negative energy states should have produced an electric field of infinite divergence according to Maxwell's equation

$$\text{div}\mathbf{E} = 4\pi\rho \tag{4.26}$$

Thus, Dirac interpreted the charge density $\rho$ in the above equation as a departure from 'the normal state of electrification of the world', which according to the hole picture would consist in all the negative energy states being occupied and all the positive energy states unoccupied. But since this ideal state could not be realized, the actual charge density $\rho$ was regarded as the sum of the charges $-e$ corresponding to the actually occupied positive energy states, and the charges $e$ corresponding to the actually unoccupied negative energy states (holes). Even granted the plausibility of this solution, there was still the problem of explaining the different masses of the electrons and protons. Furthermore, another difficulty lurked around the corner. Oppenheimer, and independently Dirac,[71] pointed out that electrons and protons would annihilate each other with the result that the mean lifetime for matter would be of the order of $10^{-10}$ seconds. When Weyl finally proved that the masses of the positive- and negative-charged electrons had to be identical, Dirac published a paper in which the identification of the holes with protons was abandoned and the concept of antiparticle was explicitly introduced:

> A hole, if there were one, would be a new kind of particle unknown to experimental physics, having the same mass and opposite charge to an electron. We may call such a particle an anti-electron. We should not expect to find any of them in nature on account of their rapid rate of re-combination with electrons.[72]

With almost prophetic words, Dirac anticipated – via the hole theory – the discovery of the positron, detected two years later by Anderson[73] in photographs of cosmic ray tracks in a Wilson cloud chamber. The discovery of the positron vindicated the hole theory, which in 1931 Dirac himself was about to give up as a sick idea because of the impossibility of making any further progress with it.[74] Moreover, it paved the way to the

---

[71] Oppenheimer (1930) and Dirac (1930b).  [72] Dirac (1931), p. 61.  [73] Anderson (1933).
[74] See Kuhn's interview of Dirac (14 May 1963) in AHQP.

elaboration of a theory of electrons and positrons perfectly symmetrical under their exchange. Particles and antiparticles could now be treated as the quanta of a field that can be created or annihilated. The interaction between matter and radiation could accordingly be treated as the interaction between quantized matter fields and the radiation fields, where matter fields can follow either Bose–Einstein or Fermi–Dirac statistics, respectively. The interaction between the electron as a quantized matter field and the quantized Maxwell electromagnetic field was explored in two pioneering papers of Heisenberg and Pauli[75] that set the stage for a dynamics of coupled quantized Dirac fields. The corresponding quantization of bosonic fields was investigated by Pauli and Weisskopf a few years later.

## 4.7 Pauli against the hole theory: the Pauli–Weisskopf 'anti-Dirac' paper

Since its inception, Pauli had strongly opposed Dirac's hole picture as unphysical and counterintuitive, to the extent of welcoming Elsasser's hypothesis of the positron as an integral spin (or null spin) particle following Bose–Einstein statistics instead of Fermi–Dirac statistics.[76] But Elsasser's hypothesis was soon refuted by the experimental result of Blackett and Occhialini at the Cavendish Laboratory:[77] positrons were indeed half-integral spin particles obeying Fermi–Dirac statistics as the hole theory predicted.

Pauli was still unconvinced. He noticed that if positrons were Fermi particles with half-integral spin, it would be possible neither for a proton to decay into a neutron and a positron, nor for a neutron to decay into a proton and an electron in the so-called β decay, on pain of either violating the conservation of angular momentum or ascribing integral or null spin to the neutron.[78] As Pauli suggested, the conservation of angular momentum could be restored in β decay only by postulating

---

[75] Heisenberg and Pauli (1929), (1930).
[76] 'Elsasser has first had the idea that the positive electron could have Bose statistics ... The fact that this point of view is so opposite to Dirac's hole theory, speaks only in its favour.' Pauli to Peierls, 22 May 1933. In Pauli (1985), p. 163.
[77] Blackett and Occhialini (1933).
[78] Since the hole theory postulated a perfect symmetry between electrons and positrons, both processes n → p, e⁻ and p → n, e⁺ were forbidden by the conservation of angular momentum. Given the spin-1/2 of the proton and of the electron, one would expect the neutron's spin to be either 1 (if the electron's and proton's spins were parallel) or zero (if they were antiparallel). But neutrons have spin-1/2. Then, it seemed that in β decay a fraction of the angular momentum of the order $1/2(h/2\pi)$ was not conserved.

a new kind of particle with mass equal or inferior to the mass of the electron and null electric charge: the neutrino.[79] Indeed, as Pauli soon realized 'the hole question and the neutrino question let themselves be solved only together.'[80]

Fermi's theory of β decay[81] provided an answer to both and constituted a historically important step towards a gradual abandonment of the hole theory. In Fermi's theory, electrons were emitted in β decay 'out of nothing' so to speak, i.e. without assuming a hypothetical negative energy sea from which they would jump by leaving behind a hole. If the hole picture had been right, one would have expected the production of a proton, an electron *and a positron* (rather than a neutrino) in β decay; furthermore, one would have expected the total number of electrons to remain constant. But neither was the case. Bloch informed Pauli about Fermi's theory, which he greeted as 'grist to my mill!'.[82] Nevertheless, the hole theory remained an important heuristic tool in quantum field theory.

In 1933 Victor Weisskopf was appointed assistant to Pauli in Zurich. This was the beginning of an intense collaboration that led the two physicists to the quantization of the relativistic Klein–Gordon equation and also put Pauli on the right track towards the spin–statistics theorem. Weisskopf recalled this crucial episode many years later:

> At that time ... the hole theory of the filled vacuum was still the accepted way of dealing with positrons. Pauli called our work the 'anti-Dirac paper'. He considered it a weapon in the fight against the filled vacuum that he never liked. We thought that this theory only served the purpose of a non-realistic example of a theory that contained all the advantages of the hole theory without the necessity

---

[79] Pauli originally postulated the existence of this new particle in 1931 to interpret the spectrum of β rays. At that time two rival hypotheses were available to explain this phenomenon: Bohr claimed that in β decay energy was not strictly conserved; Pauli, on the other hand, contended that energy was strictly conserved and the apparent violation was rather due to the emission of another type of particle, not yet observed, having null electric charge and half-integral spin so that the sum of the energies of this new particle and of the electron was constant. The new particle that Pauli originally called 'neutron', was called 'neutrino' by Fermi, after Chadwick's experimental discovery of the neutron in 1932.

[80] Pauli to Heisenberg, 11 November 1933. In Pauli (1985), p. 226.

[81] Fermi (1934). Fermi's debt to Pauli was stated at the beginning of the article: 'Following Pauli's proposal, we can assume that in β decay not only one electron, but also a new particle, the so-called "neutrino" (with mass equal or smaller than the electron's; no electric charge) is emitted. We will ground the present theory on the hypothesis of neutrino.' *Ibid.*, p. 161. The experimental detection of the new particle was possible only twenty years after Fermi's theory, thanks to the use of nuclear reactors. Indeed it was discovered that a neutron decays into a proton, an electron and an antineutrino ($n \to p + e^- + \bar{\nu}$), but because of the very small cross-section of this reaction it was very difficult to detect the antineutrino until Cowan and Reines's experiment (1956).

[82] Pauli to Heisenberg, 7 January 1934. In Pauli (1985), p. 248.

of filling the vacuum. We had no idea that the world of particles would abound with spin-zero entities a quarter of a century later. That was the reason why we published it in the venerable but not widely read *Helvetica Physica Acta*.[83]

The Pauli–Weisskopf 'anti-Dirac' paper represented Pauli's official assault on the hole theory following many previous informal attacks, as Pauli's scientific correspondence during 1933–4 testifies. Apart from β decay, one of the main problems with the hole theory was vacuum polarization, which Dirac[84] addressed at the Solvay Congress in October 1933, and whose solution Pauli judged unsatisfactory. Dirac depicted the vacuum state as a state in which no positive energy state is occupied but all the negative energy states are occupied so that no net field is produced. A field would be produced only by an occupied positive energy state (with charge $-e$) and an unoccupied negative energy state (a hole with charge $+e$). Using the Hartree–Fock approximation method, Dirac found a rule for subtracting the ideal vacuum state from the physically actual one, so as to obtain a finite difference that could figure in Maxwell's equation.

The programmatic intent of Dirac's 'subtraction physics' was to subtract the infinite vacuum distribution from the actual infinite distribution, so as to obtain a finite difference between these two infinities that was interpreted physically as the polarization of the vacuum by an external electrostatic field. It was supposed that if an electron were introduced in the vacuum state, its electrostatic field would disturb the vacuum distribution and induce an extra charge density that would neutralize a fraction of the 'true' charge of the electron. The measured charge of the electron would then amount to the 'true' charge minus a fraction of the order $-(e^2/hc)\rho$ or $-(1/137)\rho$, neutralized through vacuum polarization.

From late 1933 through spring 1934, Pauli collaborated with Heisenberg to inspect the credentials of Dirac's subtraction physics. The two main obstacles were (i) a clear understanding of the fine structure constant $\alpha = e^2/hc = 1/137$ involved in the charge renormalization, and (ii) the infinite self-energy of the electron: the energy of the electromagnetic field generated by the electron (electrostatic self-energy) together with the energy of the interaction of the electron with this field (electrodynamic self-energy) was infinite. As Pauli wrote to Heisenberg on 6 February 1934:

> Thus is Dirac's construal of laws of nature set upon Mount Sinai. All is expressed mathematically very elegantly. But physically I am not at all

---

[83] Weisskopf (1983), p. 70.   [84] Dirac (1934a).

convinced. What use is it for vacuum polarization to be finite, if the self-energy is infinite and the frequency of pair production incorrect? I am presently very disgusted with the hole theory.[85]

Indeed, the frequency of pair production calculated by Oppenheimer and Plesset[86] turned out to be wrong. And Dirac's subtraction method, further developed in a new paper,[87] left the problem of the infinite self-energy unsolved. Nonetheless, Dirac's work set the guidelines for future research and stimulated in different ways Pauli, Heisenberg, and Weisskopf. Pauli, sticking to his original scepticism, invited Heisenberg and Weisskopf to write a *Drei-männer-Arbeit* against Dirac.[88] He sketched a draft entitled 'Contribution to the theory of electrons and positrons' [*Beiträge zur Theorie der Elektronen und Positronen*].[89]

But the three-man paper never came to light because the collaboration between Pauli and Heisenberg soon terminated. Heisenberg became involved in Dirac's subtraction physics: his paper in June 1934[90] was meant to be a continuation and an extension of Dirac's project of removing infinities. Pauli and Weisskopf carried alone their two-pronged assault on Dirac's theory: in March, Weisskopf submitted to *Zeitschrift für Physik* a paper on the problem of the electron's self-energy,[91] while in July the Pauli–Weisskopf 'anti-Dirac' paper was submitted to *Helvetica Physica Acta*.[92]

Since the troublesome infinite quantities entered Dirac's theory directly through field quantization, Pauli thought that only by abandoning Dirac's theory was it possible to avoid them. Hence Pauli began to work on the rival Klein–Gordon equation, under Weisskopf's influence.[93] In a letter to Heisenberg on 14 June 1934 Pauli announced enthusiastically:

---

[85] In Pauli (1985), p. 276.   [86] Oppenheimer and Plesset (1933).   [87] Dirac (1934b).
[88] Pauli's letter to Heisenberg, 17 February 1934. In Pauli (1985).
[89] Manuscript draft in AHQP (45, 16), pp. 9. Reprinted in Pauli (1985), pp. 294–300.
[90] Heisenberg (1934).   [91] Weisskopf (1934).   [92] Pauli and Weisskopf (1934).
[93] As Weisskopf (1983), p. 69, later recalled, 'during that time I studied the properties of the Klein–Gordon equation for charged scalar particles. It seemed to me a rather academic activity, because no scalar particle was known at that time. In that theory ... it seemed possible that, under the influence of external magnetic fields, the total intensity ... might change in time, although the total charge remains conserved. It smelled of a creation and annihilation process of oppositely charged particles. I was unable to develop the problem further because I was not accustomed to use creation and destruction operators. I went to Pauli for help ... When he finally got caught up with the problem, it attracted him because he immediately saw that the quantized Klein–Gordon equation gives rise to particles and antiparticles and to pair creation and annihilation processes without introducing a vacuum full of particles.'

## 4.7 Pauli against the hole theory

> The application of our old formalism of field quantization [Heisenberg and Pauli (1929), (1930)] to the [Klein–Gordon] theory leads *without any further hypothesis*\* (without the 'hole' idea, without limiting acrobatics, without subtraction physics!) to the existence of positrons and to processes of pair production with easily calculable frequency!
>
> \*There is (after field quantization) automatically only positive energy! Everything is relativistically and gauge invariant![94]

In fact, despite the 'anti-Dirac' intentions, what Pauli and Weisskopf obtained was not a theory of electrons and positrons, but rather a theory of spin-0 particles that was perfectly symmetric with respect to positive and negative charges. But at the time, no spin-0 particle had yet been experimentally discovered and the Pauli–Weisskopf theory seemed more a formal exercise against Dirac than a genuine physical theory. Indeed, Pauli confessed that the 'spin does not enter in a relativistically invariant manner. Thus, practically speaking, we cannot proceed much further with this curiosity but it has made me happy that I can again cast aspersion on my old enemy, the Dirac theory of the spinning electron.'[95] Pauli and Weisskopf tried to force the spin into the quantization of the Klein–Gordon equation on pain of revoking the gauge and relativistic invariance of the theory as well as the positive definite energy density. This difficulty turned out to be heuristically fruitful: it cleared the ground for the following investigation of a possible spin–statistics connection. As Pauli noticed:

> Our theory can be implemented only with Bose–Einstein statistics, because here a necessary connection begins to dawn between spin and statistics.[96]

The second quantization of the Klein–Gordon equation revealed that spin-0 particles could not be fermions. Pauli and Weisskopf remarked that 'it is not possible to carry through a scalar relativistic wave theory consistently for particles obeying the exclusion principle.'[97] The search for a spin–statistics connection for particles obeying the exclusion principle proved the main stumbling-block towards the spin–statistics theorem.[98] Indeed, Pauli's first proof of the theorem came out as incomplete.

---

[94] In Pauli (1985), p. 329. Emphasis in the original.  [95] *Ibid.*, p. 329
[96] Pauli's letter to Heisenberg, 7 November 1934. In Pauli (1985), p. 361.
[97] Pauli and Weisskopf (1934), p. 722.
[98] For a comprehensive account of the history of the theorem and an excellent anthology of the various proofs, see Duck and Sudarshan (1997).

## 4.8 Pauli's first proof of the spin–statistics theorem

Pauli arrived at the spin–statistics connection in the endeavour to find for particle creation and annihilation a relativistic theory alternative to Dirac's hole theory. Prompted by the results of the 'anti-Dirac' paper, in 1936 Pauli further explored the possibility of developing a relativistic scalar theory for particles without spin but obeying the exclusion principle, a theory that – in Pauli's intention – should imply 'the production of pairs and which may be more satisfactory from the point of view of logic and pedagogy than the hole theory of Dirac'.[99] However, such a theory turned out to be unavailable as a closer analysis of the formal apparatus already developed in 1934 revealed.

Pauli followed the same procedure of second quantization of the Klein–Gordon equation already deployed in the Pauli–Weisskopf paper. He then introduced a crucial property: the charge density $\rho$ at two spatially separated points $(x - x')$ was required to commute

$$[\rho(x), \rho(x')]_{-} = 0 \qquad (4.27)$$

The vanishing of the commutator at space-like distances is an important relativistic requirement known as microcausality: this is a locality condition that forbids measurements of a physical property (in this case, the charge density) at two points with a space-like separation (i.e. that cannot get into contact through light signals) influencing each other.

Pauli then proved that it was impossible to retain this fundamental property if spin-0 particles were quantized according to the exclusion principle. In other words, the charge density $\rho(x)$ and $\rho(x')$ commuted for space–like distances $(x - x')$ if and only if spin-0 particles were quantized according to Bose–Einstein statistics. In this way, by appealing to microcausality, Pauli proved the spin-statistics connection for spin-0 particles. But he still had not proved the corresponding case for half-integral spin particles. Solving the more complicated case for particles obeying the exclusion principle was Pauli's goal in the following four years.

## 4.9 Pauli's final proof of the spin–statistics theorem

Pauli's 1936 proof set forth the guidelines for Markus Fierz's following proof in 1939.[100] At that time, Fierz was Pauli's assistant at the ETH of

---

[99] Pauli (1936), p. 138.   [100] Fierz (1939).

## 4.9 Pauli's final proof of the spin–statistics theorem

Zurich; he elaborated a complete proof of the spin–statistics theorem following Pauli's anti-Dirac lines. Fierz's proof for spin-0 particles was based on the same relativistic requirement of microcausality that Pauli had already introduced for the charge density. The real novelty consisted in the proof for half-integral spin particles, which Fierz derived from a further postulate: the positivity of the energy. The fields corresponding to half-integral spin particles were associated with spinors. Fierz introduced anticommutation relations for the quantized Fourier coefficients of the half-integral spin fields so as to satisfy the Pauli exclusion principle. This in turn made it possible, via the hole theory, to make the energy positive. The roundabout route via Dirac's hole theory was an inevitable step in order to make the total energy positive definite. Thus, half-integral spin particles turned out to be connected with Fermi–Dirac statistics because only the exclusion principle, via the hole theory, guaranteed a positive definite energy.

The necessity of resorting to both microcausality and positive energy was further stressed by Pauli in his criticism of Belinfante's[101] more or less contemporary proof: in a joint paper, Pauli and Belinfante[102] remarked that causal commutations at space-like distances implied Bose statistics for spin-0 particles, whereas positive energy implied Fermi statistics for half-integral spin particles.[103] In a letter to Weisskopf on 10 March 1940,[104] Pauli launched the final attack against Dirac's *Subtraktions-Artistik* on the occasion of Dirac's reply to a recent work of Weisskopf on the self-energy of the electron.[105] In the same letter Pauli informed Weisskopf of the delayed publication of his contribution to the Solvay Congress of 1939, due to the imminent war and related practical difficulties with the French translation. He mentioned also a more recent paper, in joint authorship with Fierz,[106] where a relativistic theory for particles with arbitrary (half-integral and integral) spin was proposed in analogy with a corresponding theory of Ettore Majorana.[107] The advantage of Majorana's theory was that the particle density and the total energy came out automatically positive, by contrast with the Pauli–Fierz theory in which a positive total energy had to be postulated. On the other hand, Pauli noticed that

---

[101] Belinfante (1939).   [102] Pauli and Belinfante (1940).
[103] More precisely, Pauli and Belinfante introduced the following postulate: 'I) for these particles *there is no infinite number of states of negative energy* ... The ... postulate is necessary since otherwise particles would drop into states of lower and lower energy creating an infinite number of quanta or pairs of particles.' Pauli and Belinfante (1940), p. 181. Emphasis in the original.
[104] In Pauli (1993), p. 24.   [105] Weisskopf (1939).   [106] Fierz and Pauli (1939).
[107] Majorana (1932).

Majorana's theory fell prey to pathological solutions with imaginary mass violating microcausality.

In the meantime, the advent of World War II forced Pauli to accept an invitation as Visiting Professor from the Institute for Advanced Study at Princeton. On 31 July 1940, Pauli and his wife left Geneva for Lisbon to continue their voyage to USA, where they arrived on 24 August. On 19 August, Pauli's final proof of the spin–statistics theorem had been received by *Physical Review*.

The article 'The connection between spin and statistics' was a re-elaboration of Pauli's report at the Solvay Congress of 1939. Like Fierz, Pauli too resorted to the spinor representation of the proper Lorentz group. Microcausality was once again at work: by postulating causal commutation or anticommutation relations for spinors, Pauli concluded that for integral spin fields, the commutator between the field and its complex conjugate vanished at space-like distances, giving the right spin–statistics connection for bosons. But for half-integral spin fields, there was no *a priori* restriction of this type: both the commutator and the anticommutator were formally allowed. To restrict the possibility to the anticommutator, Pauli, like Fierz, had to resort to the requirement of positive energy, which was in turn grounded on the hole theory:

> Hence we come to the result: *For integral spin, the quantization according to the exclusion principle is not possible* ... On the other hand, it is formally possible to quantize the theory for half-integral spins according to Einstein–Bose statistics, *but ... the energy of the system would not be positive*. Since for physical reasons it is necessary to postulate this, we must apply the exclusion principle in connection with Dirac's hole theory.[108]

Ironically, Pauli ended up relying on his old enemy (the Dirac hole theory) to get the right spin–statistics connection for fermions. So, despite the remark that 'the connection between spin and statistics is one of the most important applications of the special relativity theory',[109] Pauli was not able to prove the whole theorem on the basis only of the relativistic requirement of microcausality, and he fell back on the dichotomy microcausality/positive energy.[110]

---

[108] Pauli (1940), p. 722. Emphasis in the original.   [109] *Ibid.*, p. 722.
[110] A unified treatment of integral and half-integral spin particles under the requirement only of microcausality seemed to be unavailable to Pauli's generation. And it has continued to remain as such to the eyes of the most recent generations, if we consider the deeply instilled tendency – typical of many current quantum field theory textbooks – to present the spin–statistics connection by resorting to this dichotomy. For an analysis of this textbook

The spin–statistics theorem marks the beginning of a new phase in the history of the exclusion principle. With it, it was finally proved that *any* half-integral spin particle must follow the exclusion principle and Fermi–Dirac statistics (*fermions*), while any integral spin particle must follow Bose–Einstein statistics (*bosons*).[111] Prima facie, the theorem seems to imply the dichotomy that all elementary particles divide into fermions or bosons. In fact, the spin–statistics theorem does not depend on this dichotomy, which has attained an undeserved status through the so-called symmetrization postulate, as I shall clarify in the next chapter. The aim of the theorem is only to forge a link between the kind of spin a particle has and the kind of quantum statistics it does *not* follow, without ruling out a priori the possibility of intermediate (neither symmetric nor antisymmetric) states. Indeed, the possibility of intermediate statistics was explored from the 1940s, and in the 1960s led to the development of a research programme that investigated half-integral spin particles violating Pauli's exclusion principle. I shall come back to this programme in the next chapter. In the final section of this chapter, in the light of the historical reconstruction just completed, I want to make some philosophical remarks about the nomological shift that Pauli's exclusion rule underwent with the development of the new theoretical framework.

## 4.10 How Pauli's rule gained the status of a scientific principle

Pauli's rule became a scientific principle, and indeed a fundamental one when a new theoretical framework, into which the rule was built from the ground up, came to be articulated. In this chapter I have reconstructed the main steps of this process: the development of quantum statistics, which redefined Pauli's rule as a prescription of antisymmetrizing the wave function for an assembly of indistinguishable particles; the electron's spin recast in terms of Pauli's spin matrices with their subsequent group-theoretic derivation; the further reinterpretation of Pauli's principle in terms of anti-commutation relations for particle creation and annihilation operators in quantum field theory, from which the spin–statistics theorem was proved in

---

tradition and its comparison to Weinberg's (1964) proof, which on the contrary resorts only to microcausality to prove the spin–statistics connection for both integral and half-integral particles, see Massimi and Redhead (2003).

[111] Let me say just as a clarifying remark that the spin–statistics theorem fails in one and two spatial dimensions. Two-dimensional entities, the so-called anyons, do not obey the spin–statistics theorem and this has applications in surface film phenomena and superconductivity, see Wilczek (1990).

1940. With this theorem, the transition from the status of a phenomenological rule for electronic distribution as of 1924 to that of a scientific principle whose nomological validity extends to all half-integral spin particles, was completed.

It was via this incorporation into a growing theoretical framework that Pauli's rule could transcend its humble origins and enlarge its nomological scope. Pauli's principle should not then be regarded as a coordinating principle *à la* Reichenbach: it did not coordinate a pre-existing mathematical structure with the proper physical part of the new quantum mechanics. Nor, hence, could it be a constitutive a priori principle – albeit a relativized one – *à la* Friedman: as we saw in Chapter 1, constitutive a priori principles lay the conditions of possibility of a scientific framework by accomplishing such a coordinating task.

Rather, following the constitutive–regulative distinction in the Kantian tradition presented in Section 1.4, I want to suggest that the necessity and nomological strength of Pauli's principle is the result of the *systematization* of this increasing body of quantum mechanical knowledge where systematization must be understood in Kantian terms as a regulative principle, and, as such, as open-ended and never fully attainable. This process of systematization involved the gradual reorganization and subsuming of lower-order concepts into higher-order, i.e. more abstract, ones (e.g. from the semi-classical spinning electron to Pauli's spin matrices and their group-theoretic derivation), as well as subsuming phenomena into a growing scientific architectonic (from the non-relativistic quantum mechanics of the magnetic electron to relativistic quantum field theory) from which they could naturally follow.

The reformulation of Pauli's *Verbot* in terms of Fermi–Dirac statistics played a major role in this process of systematization. Via this reformulation as a prescription to antisymmetrize the wave function, Pauli's *Verbot* could transcend its humble origins in atomic spectroscopy and be embedded in the new quantum theory. Later, the principle came to forbid bosonic quantization for Dirac fields; as such, via the spin–statistics theorem, its scope of applicability could be extended to *all* half-integral spin particles. In this way, more and more phenomena were brought under the prescriptive domain of Pauli's *Verbot*; at the same time, via this same process, necessity and nomological strength accrued to Pauli's rule, to the extent of transforming it into a scientific principle, and a fundamental one indeed, of the new framework. A dynamic Kantianism *à la* Buchdahl/Cassirer, as foreshadowed in Chapter 1, is therefore congenial to the history of the exclusion principle. It accounts for the necessity and lawlikeness of the principle as a function of its role in the systematization of quantum mechanical knowledge, while leaving intact the empirical (and experimentally testable) nature of the principle.

The articulation of the new framework around a phenomenological law deriving from the old quantum theory reinforces the image suggested in Chapter 3. Against Kuhnian incommensurability, I there argued for a natural evolution of scientific concepts and nomic generalizations during inter-lexicon shifts: the revolutionary new concept of the electron's spin and the associated nomic generalization (Pauli's exclusion rule) followed from phenomena and theoretical assumptions of the old quantum theory via a process of gradual conceptual and theoretical revision. In this chapter, we have further seen how they were built into the new framework from the ground up in a way that affected the same nomological status of Pauli's rule. Once again, the passage from the old quantum theory to the establishment of the new quantum theory was smoother than the Kuhnian picture may suggest.

The new framework allowed a rigorous formal derivation of all those anomalous phenomena that the old quantum theory could at best accommodate via suitably tailored rules of thumb. The 'duplexity phenomena' and Thomas's relativistic correction put in 'by hand' in the spinning electron model finally followed from the Dirac relativistic equation. Complex spectra, selection rules, and the Zeeman effect, which all appeared as a hopeless conundrum in the early 1920s, naturally followed from relativistic quantum mechanics. The new framework was able to explain why the semi-classical, faulty spinning electron model was nonetheless empirically successful; or, as Dirac put it, why there was a great deal of truth in the spinning electron model at least as a first approximation. The revolutionary transition became *retrospectively intelligible*. While the prospective intelligibility was grounded on the stepwise transformation of the old quantum theory (Chapter 3), the retrospective intelligibility was warranted by the ability of relativistic quantum mechanics to derive rigorously the results previously obtained a posteriori via a mixture of numerology and empirically introduced factors.

But it was not only old evidence that could be derived from the new framework. New evidence too became accessible, undreamt-of evidence that had not been used in the building-up of the framework itself. Novel phenomena were found, whose history is intertwined with the history of the electron's spin and Pauli's principle. We saw in the course of this chapter the experimental discovery of the positron in 1932, anticipated by Dirac's hole theory; and of the neutrino predicted by Pauli and Fermi's theory of β decay as early as 1930. More novel phenomena were just around the corner, and it was the exclusion principle in its enlarged nomological scope granted by the spin–statistics theorem that paved the way for their experimental detection. This required long and laborious experimental activity, which in the end strengthened the exclusion principle and, for the first time, made it possible to test it. This is the history of coloured quarks, paraparticles, and experimental limits to possible violations of the exclusion principle that I shall present in the next chapter.

# 5

# The exclusion principle opens up new avenues: from the eightfold way to quantum chromodynamics

This chapter explores the heuristic fruitfulness of the exclusion principle in opening up new avenues of research: namely the idea of 'coloured' quarks and the development of quantum chromodynamics (QCD) in the 1960s to 1970s. Sections 5.1 and 5.2 reconstruct the origin of the quark theory from Gell-Mann's so-called 'eightfold way' for elementary particles. The experimental discovery of the $\Omega^-$ particle confirmed the validity of Gell-Mann's model, but it also provided negative evidence against quarks obeying the exclusion principle. Two alternative research programmes emerged in the 1960s to deal with this piece of negative evidence (Section 5.3): the first programme (Section 5.3.1) rejected the strict validity of the exclusion principle and explored the possibility that quarks obeyed parastatistics; the second (Section 5.3.2) retained the exclusion principle and reconciled it with negative evidence by introducing a further degree of freedom for quarks ('colour'). The Duhem–Quine thesis seems to loom on the horizon (Section 5.4): the choice as to whether questioning the principle or introducing an auxiliary assumption to reconcile it with negative evidence seems to be underdetermined by evidence. However, I shall argue that it was exactly via the development of these two rival research programmes that the exclusion principle came to be validated, and that there was a rationale for retaining the principle despite prima facie recalcitrant evidence.

## 5.1 Introduction

As we have seen in Chapter 4, soon after 1924 Pauli's exclusion rule was incorporated into the growing quantum mechanical framework, where its role came to be redefined and its nomological scope extended. Thanks to the role it played in the systematization of this growing body of quantum mechanical knowledge – once it had been reformulated as a prescription to antisymmetrize the wave function – Pauli's *Verbot* could pass from the status of a tentative phenomenological law in the old quantum theory to that of a scientific principle of the new quantum theory. Until 1930 its nomological scope had been confined to electrons and to their antiparticles. With the spin–statistics theorem, it was extended to *any* half-integral spin particle.

In the following two decades, the family of half-integral spin particles grew to include brand new particles beside electrons, positrons, neutrons, and protons. In 1956 Cowan and Reines detected the neutrino, whose existence Pauli had predicted in 1931 to explain β decay. In 1962 Lederman, Schwartz, and Steinberger distinguished the muonic from the electronic neutrino. Another family of particles with half-integral spin and mass greater than the mass of nucleons was soon discovered, the so-called hyperons. They included a neutral particle called $\Lambda^0$, a group of three particles with charge +1, 0, −1 ($\Sigma^+$, $\Sigma^0$, $\Sigma^-$) and a pair of particles with charge −1 and 0 ($\Xi^-$, $\Xi^0$). In 1964 another half-integral spin particle was discovered (the $\Omega^-$), which will play a major role in the story I am going to tell in this chapter. As we shall see, the discovery of the $\Omega^-$ confirmed a phenomenological model that Murray Gell-Mann had proposed at the beginning of the 1960s to classify elementary particles (the so-called 'eightfold way'): this model postulated the existence of quarks as the basic constituents of hadrons.[1] Since quarks are themselves half-integral spin particles, in the light of the spin–statistics theorem they too were expected to obey the exclusion principle: there could not be two 'equivalent' quarks. But experimental evidence fell short of this expectation: some baryons (namely, the $\Omega^-$) turned out to be made of three equivalent quarks (three s quarks). Furthermore, baryon spectra had been successfully classified at the cost of assuming that the three quarks composing each baryon were in symmetric states, in contrast with the antisymmetric states required by the exclusion principle.

Two options were available to deal with these pieces of recalcitrant evidence for Pauli's principle within the quark theory: either to violate the exclusion principle and allow half-integral quarks to be in non-antisymmetric

---

[1] Hadrons include mesons and baryons. Mesons are particles with intermediate mass such as pions ($\pi^+$, $\pi^0$, $\pi^-$), kaons ($K^+$, $K^0$, $K^-$), and the $\eta$ meson. Baryons are particles with high mass, e.g. nucleons and hyperons, although the current classification does not necessarily follow this terminological distinction (e.g. the B is a meson despite its very high mass). A more precise formulation of the meson/baryon distinction is in terms of their quark composition. As will be clarified later, mesons consist of a quark–antiquark pair; baryons consist of a quark triplet.

states; or to maintain the exclusion principle by introducing a further degree of freedom for quarks. The first option led Messiah and Greenberg to the elaboration of parastatistics (Section 5.3.1); the second led to coloured quarks (Section 5.3.2). The very idea of coloured quarks was introduced in an attempt to maintain the exclusion principle in its enlarged nomological scope established by the spin–statistics theorem in the face of some negative evidence in quark theory.

This historical episode evokes the Duhem–Quine dilemma. Given the holistic nature of scientific knowledge, if some negative evidence is discovered, we are in no position to know whether the blame for it lies with the main theory or with one of the auxiliary assumptions conjoined with it. Hence the underdetermination of theory by evidence: the choice as to whether rejecting the main theory or simply changing one or more of the auxiliary assumptions is underdetermined by evidence, i.e. empirical evidence by itself is unable to determine which option we should go for. The existence of two rival research programmes (parastatistics and QCD), respectively based on the rejection versus acceptance of the exclusion principle within the quark theory, nicely illustrates this problem.

However, epistemological holism need not be a royal road to underdetermination. On the contrary, it can fruitfully lead to a 'holistic' validation of a threatened scientific principle. The exclusion principle, built into QCD from the ground up, has disclosed undreamt-of phenomena widely confirmed by experimental evidence in the past forty years, while the rival parastatistics programme – in its subsequent versions – has opened up the unprecedented possibility of experimentally testing the principle by predicting Pauli-violating states. In the final section of this chapter, I am going to show how the validation of the exclusion principle passed through the development of these two rival research programmes, and how the rationale for retaining the principle resides ultimately in the empirical support that accrued to it.

## 5.2 From the eightfold way to quarks

The discovery of new particles in the 1950s brought with it the introduction of new physical properties with related conservation laws.[2] For instance, it was observed that the creation probability of kaons and hyperons in the

---

[2] For an excellent reconstruction of the experimental discovery of these particles and of the history of the eightfold way, see Ne'eman and Kirsh (1986), on which I draw for this section.

collision of two protons was very high: this indicated that the strong interaction was responsible for this creation process. On the other hand, since the decay of kaons and hyperons was very slow, the weak interaction seemed to be at work in the annihilation process. Furthermore, it was observed that kaons and hyperons were always created and annihilated as a pair (e.g. $\Lambda^0 + K^0$; $K^+ + \Sigma^+$; $K^- + K^0$). This phenomenon was similar to electron–positron creation from $\gamma$ ray scattering ($\gamma \rightarrow e^+ + e^-$); and since electron–positron annihilation satisfied the conservation of charge, it was hypothesized that a similar conservation law could be at work for kaons and hyperons. There must be a property of kaons and hyperons that was conserved in the strong interactions and violated in the weak interactions. This new property was called *strangeness* and it was associated with a new quantum number $S$. On the basis of theoretical considerations, the hyperon $\Lambda^0$ was assigned $S = -1$; the kaons $K^0$ and $K^+$ were assigned $S = +1$; the $\Sigma$ hyperons $S = -1$; the $\Xi$ hyperons $S = -2$ and, finally, a hypothetical not then discovered hyperon, the $\Omega^-$, was assigned $S = -3$. The introduction of the new property allowed the classification of hadrons into groups (called multiplets) having the *same* strangeness, baryon number,[3] and spin: (p, n); ($K^+$, $K^0$); ($\Xi^0$, $\Xi^-$); ($\Sigma^+$, $\Sigma^0$, $\Sigma^-$); and the singlets ($\Lambda^0$), ($\Omega^-$).

This classification suggested that the particles composing a multiplet could be taken as different aspects of one and the same entity. This idea can be traced back to Heisenberg, who in 1932 proposed regarding protons and neutrons as two states of the same particle (the nucleon) 'spinning' in opposite directions in a hypothetical three-dimensional space so that the nucleus of an atom was invariant under a 360° rotation in any direction (in group theory terminology, the nucleus was invariant under SU(2) transformations). The SU(2) transformations are $2 \times 2$ matrices corresponding to the two possible states of any nucleon; but the SU(2) group can also be represented in $n$ dimensions, corresponding to the eventual $n$ states of the entity. The step from this mathematical property of the SU(2) group to the introduction of the isospin was easy.

The particles of any multiplet were taken to be different 'spin' states – so to speak – of the same entity 'spinning' in the $x$-, $y$-, $z$-directions of an abstract three-dimensional space, the so-called *isospin I*. The projection of the isospin on the $z$-direction was denoted by the new quantum number $I_3$. Given a multiplet of $n$ particles, the isospin of each group is given by

---

[3] The baryon number (denoted by the quantum number $A$) is associated with another conservation law: the number of baryons minus the number of antibaryons must be constant. Protons, neutrons, hyperons $\Sigma$ and $\Lambda$ were all assigned $A = 1$, while their respective antiparticles were assigned $A = -1$.

## 5.2 From the eightfold way to quarks

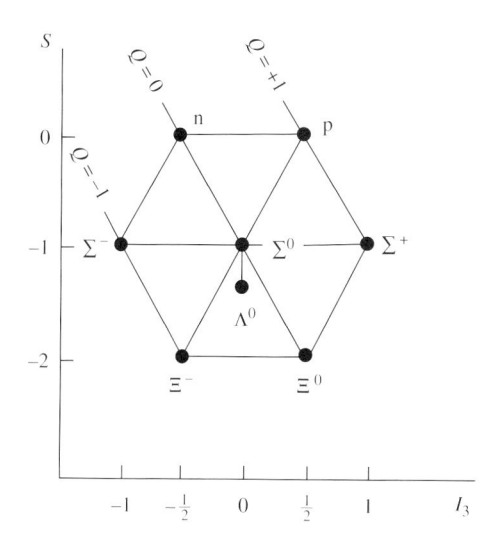

Fig. 5.1 The octet of spin-1/2 baryons according to the eightfold way.

*Source:* Y. Ne'eman and Y. Kirsh (1986) *The Particle Hunters* (Cambridge: Cambridge University Press). Reproduced with permission of Cambridge University Press.

$I = (n-1)/2$. Accordingly, the isospin of doublets such as (p, n), ($K^+$, $K^0$), ($\Xi^0$, $\Xi^-$) is $I = 1/2$ (with $I_3 = 1/2$ for protons and $I_3 = -1/2$ for neutrons); the isospin of triplets such as ($\Sigma^+$, $\Sigma^0$, $\Sigma^-$) is $I = 1$ (with $I_3$ equal to the charge of each particle), while the isospin of singlets ($\Lambda^0$) and ($\Omega^-$) is $I = 0$ (with $I_3 = I = 0$).

In 1961, Murray Gell-Mann and Yuval Ne'eman, independently of each other, proposed unifying multiplets with the same baryon number, spin and parity into greater theoretical structures called supermultiplets.[4] The supermultiplets were represented in a system of Cartesian coordinates with the *x*-axis denoting the component of the isospin $I_3$ and the *y*-axis denoting the strangeness *S*. Each particle of a supermultiplet was accordingly represented as a point on the $I_3$–*S* plane: the octet of all semi-stable baryons with spin-1/2 (i.e. (p, n), ($\Xi^0$, $\Xi^-$), ($\Sigma^+$, $\Sigma^0$, $\Sigma^-$), ($\Lambda^0$)) turned out to compose a symmetric hexagon on the $I_3$–*S* plane (Fig. 5.1). Symmetric hexagons were similarly found for the octets of spin-0 and spin-1 mesons, while the ten-component supermultiplet of spin-3/2 baryons – known as the $\Omega^-$ decuplet – formed a triangle on the $I_3$–*S* plane (Fig. 5.2): it included a group of four resonances ($\Delta^-$, $\Delta^0$, $\Delta^+$, $\Delta^{++}$) that Fermi had discovered in 1952; a group of three resonances considered as excited states of the hyperon

---
[4] Gell-Mann (1961); Ne'eman (1961).

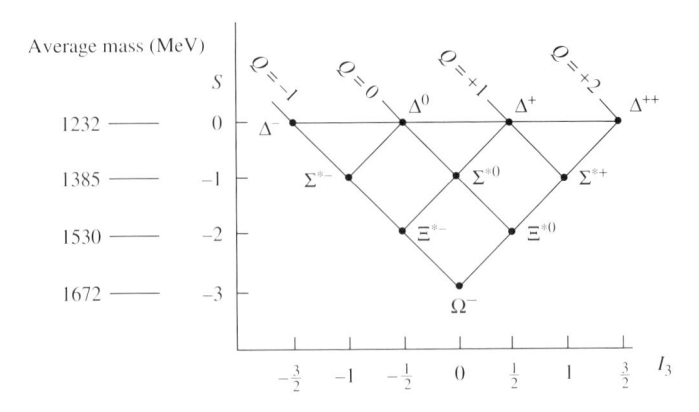

150    5 *From the eightfold way to quantum chromodynamics*

Fig. 5.2 The $\Omega$ decuplet of spin-3/2 baryons according to the eightfold way.
*Source:* Y. Ne'eman and Y. Kirsh (1986) *The Particle Hunters* (Cambridge: Cambridge University Press). Reproduced with permission of Cambridge University Press.

$\Sigma$ ($\Sigma^{*-}$, $\Sigma^{*0}$, $\Sigma^{*+}$); two resonances corresponding to the hyperon $\Xi$ ($\Xi^{*-}$, $\Xi^{*0}$), and finally the hypothetical $\Omega^-$, which was the missing vertex of the triangle and whose properties were predicted by Gell-Mann in 1962.[5]

Gell-Mann's and Ne'eman's idea of gathering particles into supermultiplets sharing some physical properties and forming symmetric figures on the $I_3 - S$ plane took the name *eightfold way*. It offered a phenomenological model of classification for elementary particles that turned out to be heuristically fruitful. The eightfold way was based on the following analogy: while the particles of each multiplet are different states of the same entity 'spinning' in different directions of the isospin space, the particles of each supermultiplet similarly are different states of the same entity 'spinning' in different directions of another abstract, complex space: the space of the so-called unitary spin.

The unitary spin has eight components (hence the name 'eightfold way') and follows the Lie algebra of a particular group of three-dimensional unitary matrices, known as SU(3). The algebra of the SU(3) group fixes specific constraints for the particles of each supermultiplet. For instance, it restricts the number of particles per supermultiplet to groups of 3, 8, 10, or 27. It also fixes specific relations between the masses and the magnetic moments of the components. It was in the light of these constraints imposed by SU(3) symmetry that Gell-Mann could predict the isospin,

---
[5] Gell-Mann (1962a).

## 5.2 From the eightfold way to quarks

strangeness, mass, and average lifetime of the $\Omega^-$ as the missing tenth particle of the decuplet of spin-3/2 baryons.

The $\Omega^-$ was experimentally discovered at the Brookhaven National Laboratory in February 1964,[6] in photographs of a chain process where accelerated protons produced a jet of kaons in a hydrogen bubble chamber. The quantum numbers of the new particle coincided with those predicted by Gell-Mann. The discovery of the $\Omega^-$ was a brilliant success for the eightfold way, but at the same time it raised a new theoretical challenge. The eightfold way was a successful phenomenological model in fixing some important constraints on particles, but it did not explain the reasons for these constraints. Just as Balmer's formula could not explain the spectral lines without Bohr's atomic theory; or, Landé's $g$ factors could not explain the magnetic anomaly without the electron's spin; similarly, the eightfold way could not explain why hadrons gathered so nicely into supermultiplets sharing some important physical properties without a further daring theoretical hypothesis: the existence of quarks.

The idea of quarks goes back to Ne'eman in 1962, but only with Gell-Mann[7] was it precisely formulated.[8] Hadrons were supposed to be built up from entities called quarks:[9] more precisely, mesons consisted of a quark–antiquark pair while baryons consisted of a triplet of quarks or antiquarks. Gell-Mann postulated three types ('flavours') of quarks: *up* (u), *down* (d), *strange* (s); their respective antiquarks were $\bar{u}, \bar{d}, \bar{s}$. In Gell-Mann's words:

> A simpler and more elegant scheme can be constructed if we allow non-integral values for the charges [$z$] ... We assign to the triplet $t$ the following properties: spin 1/2, $z = -1/3$, and baryon number 1/3. We then refer to the members $u^{2/3}$, $d^{-1/3}$, and $s^{-1/3}$ of the triplet as 'quarks' q and the members of the anti-triplet as anti-quarks $\bar{q}$. Baryons can now be constructed from quarks by using the combinations (q, q, q), (q, q, q, q, $\bar{q}$), etc., while mesons are made out of (q, $\bar{q}$), (q, q, $\bar{q}$, $\bar{q}$) etc. It is assumed that the lowest baryon configuration (q, q, q) gives just the representations 1, 8, 10 that have been observed, while the lowest meson configuration (q, $\bar{q}$) similarly gives just 1 and 8.[10]

---

[6] Barnes *et al.* (1964).   [7] Gell-Mann (1964).
[8] For the sake of clarity, it must be mentioned that a different approach to quarks was independently proposed by Zweig (1964a), (1964b). This was a more phenomenologically oriented approach that was successfully applied to hadron spectroscopy, but theoretically unsatisfactory because, among other things, the SU(6) group it used was not Lorentz invariant. For a reconstruction of Zweig's so-called 'constituent quark model', see Pickering (1984), pp. 89–108.
[9] The name 'quarks' was taken from James Joyce's *Finnegan's Wake*.
[10] Gell-Mann (1964), p. 214.

By assigning to quarks spin 1/2, fractional charge $z$ equal to 2/3 (for the u quark) and $-1/3$ (for d and s), baryon number 1/3 and strangeness $S=0$ for u and d, and $S=-1$ for the s quark, the baryon singlet, octet, and decuplet arose naturally. The combination of quarks allegedly endowed with the aforementioned properties explained all the known hadrons. It explained, for instance, the existence of a baryon ($\Omega^-$) with $S=-3$ and charge $-1$, while it excluded a baryon with $S=-3$ and charge 0 or $+1$, which nonetheless was allowed by the SU(3) group in a supermultiplet of 27 particles.[11] The properties of the $\Omega^-$ naturally followed from its being made out of three s quarks, each with strangeness $S=-1$ and fractional charge $z=-1/3$. Similarly, it became possible to explain the properties of nucleons, under the assumption that protons were uud triplets and neutrons were udd triplets, as well as those of the hyperons ($\Lambda^0=$ uds, $\Sigma^+=$ uus). Moreover, it was possible to explain why the spin of mesons is always integral whereas the spin of baryons is always half-integral: in the former case the spins of the quark-antiquark pair can be either parallel (with total spin 1) or antiparallel (total spin 0); in the case of baryons, made out of three quarks, the resultant total spin is either 1/2 or 3/2. The baryon number could be similarly explained, and the conservation of the baryon number was reinterpreted as the conservation of the difference between the number of quarks and the number of antiquarks. Analogously, the strangeness was reinterpreted as the difference between the number of $\bar{s}$ antiquarks and the number of s quarks, and the conservation of strangeness as the impossibility for an s quark to transform into a u or d quark.

Gell-Mann grounded the hypothesis of quarks on his earlier work on current algebra.[12] This was a field-theoretic approach that treated weak interactions in terms of leptonic and hadronic weak currents, in analogy with the QED treatment of electromagnetic interaction in terms of electromagnetic currents. On this approach, quarks were the vehicles of hadronic weak currents. In 1963, Nicola Cabibbo[13] built upon Gell-Mann's approach, by considering the d and s quark states at work in weak interactions not as pure flavour eigenstates but as a mixture of the two quark-flavour states resulting from a rotation through the so-called Cabibbo angle $\theta_C$

$$d \cos \theta_C + s \sin \theta_C \tag{5.1}$$

---

[11] See Ne'eman and Kirsh (1986), Chapter 9, Section 9.4.
[12] Feynman and Gell-Mann (1958); Gell-Mann (1962b); Gell-Mann and Lévy (1960).
[13] Cabibbo (1963).

## 5.2 From the eightfold way to quarks

Thus, the three quark flavours formed a quark doublet with the 2/3-charged u quark eigenstate, and the −1/3-charged d and s quark mixture. Cabibbo's theory explained the relative strengths of the observed charged current weak decays involving d and s quarks. However, Cabibbo's theory allowed the existence of strangeness-changing weak neutral currents: e.g. the decay of the neutral kaon into a pair of muons $K^0 \to \mu^+\mu^-$, where the total charge is conserved while the strangeness $S$ drops from +1 to 0. But, strangeness-changing weak neutral currents were observed to occur very rarely, a phenomenon that was explained only in 1970 with the introduction of a fourth quark flavour ('charm'), as we shall see in Section 5.3.2.3.

Despite the remarkable heuristic power of the quark theory, it was impossible to detect quarks in the laboratory. And, as we shall see, this was not just a temporary impossibility due to inadequate experimental apparatus: it was *in principle* impossible to detect isolated quarks because of what became known as 'colour confinement'. Moreover, the existence of quarks by itself was not very illuminating without a dynamic theory able to explain the interactions among them. And the first important step towards a dynamic theory for quarks was taken in the attempt to solve a new riddle concerning the recently discovered $\Omega^-$ particle. As mentioned above, according to the quark theory the $\Omega^-$ consisted of three s quarks all with parallel spin-1/2. Similar triplets of identical quarks with parallel spins also occurred in the resonances $\Delta^{++}$ (uuu) and $\Delta^-$ (ddd) in the $\Omega^-$ decuplet.

Since quarks have spin-1/2, according to the spin–statistics theorem they are fermions and hence obey the exclusion principle. But, surprisingly enough, analysis of the $\Omega^-$ revealed that there existed three 'equivalent' s quarks with parallel spins. Furthermore, the experimental analysis of baryon spectra required that the quark combined space and spin wave functions were symmetric under the interchange of any two quarks of the same flavour. An antisymmetric wave function (as required by the exclusion principle) would have led to baryon spectra incompatible with those experimentally observed.[14] For the first time the exclusion principle, whose strict validity in atomic and nuclear physics was by then well-established, seemed to face negative evidence. This conundrum led to a serendipitous

---

[14] It is important to clarify that this contrast with Fermi–Dirac statistics originally emerged within Zweig's 'constituent quark model' (see Footnote 8) as a phenomenological model for baryon spectra. However, since the symmetry group of the 'constituent quark model' was related by a suitable transformation to the symmetry group of Gell-Mann's 'current quark model', the discovery that the 'constituent quarks' seemed to obey peculiar statistics posed the same conundrum also for 'current quarks'. See on this point Fritzsch and Gell-Mann (1972).

154   *5 From the eightfold way to quantum chromodynamics*

theoretical result. It smoothed the path to two rival research programmes. The first programme tried to solve the tension between the quark theory and the exclusion principle by revoking the strict validity of the exclusion principle for quarks and by assuming that quarks were not fermions but parafermions. The second programme, which interestingly stemmed from the first, explored a new way of reconciling the exclusion principle with recalcitrant evidence by introducing a further degree of freedom for quarks, the colour.

## 5.3 Revoking or retaining the exclusion principle?

### 5.3.1 Revoking the strict validity of the exclusion principle: quarks as parafermions

A possible way out of the impasse consisted in revoking the strict validity of the exclusion principle for quarks. O. W. Greenberg at the University of Maryland opted for this solution: quarks would obey a quantum statistics intermediate between Fermi–Dirac and Bose–Einstein (parastatistics).

Parastatistics had been explored since 1940. While Pauli was elaborating the final proof of the spin–statistics theorem, other physicists were looking for quantum statistics alternative to Fermi–Dirac and Bose–Einstein. Indeed, the spin–statistics theorem left open the question as to whether intermediate statistics were possible. In 1940 Gentile[15] obtained a generalized intermediate statistics that degenerated into Fermi–Dirac or Bose–Einstein whenever the occupation number was equal to 1 or to infinity. But intermediate statistics soon stood condemned of violating the basic quantum mechanical assumption that every physically distinguishable state must correspond to a unique ray in the Hilbert space, picked out by a vector $|\psi\rangle$.[16] In the following years this problem was overcome by dispensing with this assumption.[17]

In 1964, Messiah and Greenberg[18] showed that the fermion/boson dichotomy established not by the spin–statistics theorem, but rather by the symmetrization postulate, did not follow from the much weaker indistinguishability postulate. The symmetrization postulate (SP) required all

---

[15] Gentile (1940).   [16] ter Haar (1952).
[17] For details about the history of parastatistics, see French (1984), pp. 250–5; French (1995), pp. 88–91.
[18] Messiah and Greenberg (1964a).

## 5.3 Revoking or retaining the exclusion principle

physical states to be either symmetric or antisymmetric. The indistinguishability postulate (**IP**), on the other hand, amounted to the weaker claim that if $|\psi\rangle$ is the vector representing the state $\Psi$ of a composite system of $n$ indistinguishable particles, then by measuring the expectation value of any observable $B$ it is not possible to distinguish $|\psi\rangle$ from any permutation $P|\psi\rangle$. IP is typically satisfied in two cases:

(a) if $|\psi\rangle$ is symmetric: $P|\psi\rangle = |\psi\rangle$, for all $P$ (bosons);
(b) if $|\psi\rangle$ is antisymmetric: $P|\psi\rangle = |\psi\rangle$ if $P$ is even, $P|\psi\rangle = -|\psi\rangle$ if $P$ is odd (fermions).

Thus, the physical states allowed by the IP are usually taken to be the Hilbert subspaces picked out by symmetric or antisymmetric vectors, hence the apparent boson/fermion dichotomy of the SP. Messiah and Greenberg showed that the IP did not entail necessarily the dichotomy of the SP. They pointed out that the SP was a consistent way of inserting the Pauli principle in the formalism of quantum mechanics, but within that formalism the group of permutation operators $P$ had irreducible representations of dimension greater than 1: the Hilbert space of a composite system of particles has subspaces (the so-called generalized rays) of dimension greater than 1, which are invariant under all permutations. These generalized rays correspond to intermediate – neither symmetric nor antisymmetric – states of higher symmetry type and obeying a quantum statistics different from both Fermi–Dirac and Bose–Einstein: parastatistics.[19] It turned out that the IP allows for symmetric, antisymmetric, and higher symmetry states too, *pace* the dichotomy apparently established by the SP. Particles following parastatistics were called paraparticles. Greenberg and Messiah's result was valid only for systems that could be described in the quantum mechanics framework, not in the quantum field theory framework, where symmetry properties of states are encoded in the algebraic relations (commutation/anticommutation relations) between field operators.

In a following article, Greenberg and Messiah[20] extended their result from quantum mechanics to quantum field theory. Parafield quantization was a generalization of the usual method of second quantization. The original idea of parafield theory goes back to H. S. Green,[21] who

---

[19] For a discussion of this literature, see French and Redhead (1988); Redhead and Teller (1992).
[20] Messiah and Greenberg (1964b).
[21] Green (1953). For later developments, see Stolt and Taylor (1970); Ohnuki and Kamefuchi (1982). For a summary see Redhead (1982).

introduced trilinear (instead of bilinear) commutation relations for paraparticle field operators. Building on Green's idea, Greenberg and Messiah introduced parafermions/parabosons, which can occur in a 'symmetric'/ 'antisymmetric' state respectively, where there was no restriction to the number of parabosons per state, but at most $p$ parafermions could occur in the same quantum state. Messiah and Greenberg invited experimental physicists to search for paraparticles in some puzzling phenomena with an unclear statistics. For instance, a new statistics of order $p = 2$ would allow double occupancy of quantum states, in contrast with the exclusion principle.

Greenberg suggested that the quark theory could be reconciled with negative evidence about the exclusion principle by assuming that quarks were parafermions of order $p = 3$.[22] In this way quarks still had separate wave functions which were symmetric under permutation of any two of them, but they also had composite wave functions which were antisymmetric under permutation with other composite states, in agreement with the spin–statistics theorem. As a parafermion of order 3, each quark was assigned three labels that – as we shall see in Section 5.3.2 – were shortly after reinterpreted (within a rival research programme) as denoting a new threefold degree of freedom for quarks, the 'colour'.

After colour was introduced and quantum chromodymanics established as a gauge theory of strong interaction based on the colour SU(3) symmetry, it became clear that Greenberg's original formulation of parastatistics was not amenable to gauging. A Yang–Mills Lagrangian could be constructed from parafields only for $p = 3$ and for the symmetry SO(3), but not for the colour SU(3) symmetry. However, as Greenberg and Nelson pointed out,[23] the possibility that colour could be gauged in a physically equivalent parafield formulation was not yet excluded. In 1983, Greenberg and Macrae[24] succeeded in formulating a locally gauge-invariant parastatistics using a complex Clifford algebra so that for the SU(3) the gauge theory of para-Fermi quarks was equivalent to quantum chromodynamics. So, while Greenberg and Messiah's original formulation of parastatistics could not be gauged, Greenberg and Macrae finally found a way of gauging parastatistics that made it equivalent to QCD. But, leaving aside this issue, in the following section I want instead to draw attention to some important experimental consequences of the parastatistics

---

[22] Greenberg (1964).     [23] Greenberg and Nelson (1977), p. 88.
[24] Greenberg and Macrae (1983).

programme, namely its opening the door to the unprecedented possibility of testing the exclusion principle.

### 5.3.1.1 From parons to experimental tests of the exclusion principle

The theoretical development of parastatistics called for experimental confirmation. A series of experiments was carried out in the following decades to test eventual violations of the exclusion principle as predicted by parastatistics. Violations were searched for in forbidden transitions in K-shell X-rays: when an electron scatters from a normal atom, the electrons are rearranged and an anomalous Pauli-violating state might appear with probability $\beta^2$ with respect to the normal state obeying the Pauli principle, with $\beta$ the parameter governing Pauli principle violation. This excited state may in turn decay into an anomalous ground K state by emitting X-rays. Thus, K-shell X-rays could provide a high-precision test for the exclusion principle. This idea had originally been explored in 1948 by Goldhaber and Scharff-Goldhaber[25] in their search for K-shell X-rays in lead: the null result of their experiment set a limit of $\leq 0.03$ for possible violations of the exclusion principle.

Forty years later, Greenberg and Mohapatra[26] further reduced this limit. By that time, paraparticles had been replaced by paronic particles or 'parons'. Parons were hypothetical particles violating the exclusion principle by only a small amount. Violations of the exclusion principle were parametrized by assuming that two identical fermions were with probability $1 - 1/2\beta^2$ in the usual antisymmetric state, and with probability $1/2\beta^2$ in an anomalous symmetric state. The parameter $\beta^2$ was bounded by about $\beta^2 < 10^{-6}$–$10^{-8}$. If the exclusion principle were violated, a fraction of order $\beta^2$ of anomalous atoms was expected to be contained in a sample of the chemical element at issue. Spectroscopic analyses were expected to reveal the presence of such anomalous atoms, whose Zeeman splitting was supposed to be proportional to appropriate powers of $\beta^2$. As Greenberg and Mohapatra noticed, the currently available spectroscopic techniques could not detect extra lines in the Zeeman splitting unless $\beta^2$ was greater than $10^{-6}$–$10^{-8}$. But two different kinds of experiments could improve the limit on $\beta^2$: (1) experiments detecting atoms in ground states prohibited by normal statistics; and (2) experiments detecting transitions prohibited by normal statistics.

In the first case, where anomalous ground states were searched for, the intermediate states were represented by a Young pattern (Fig. 5.3): the

---

[25] Goldhaber and Scharff-Goldhaber (1948).  [26] Greenberg and Mohapatra (1989).

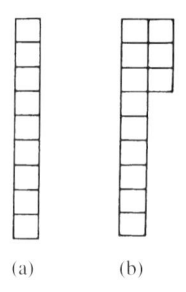

Fig. 5.3 Young patterns for an antisymmetric state (a) and for a mixed symmetry state with at most double occupancy (b).

*Source:* Reprinted with permission from O. W. Greenberg and R. N. Mohapatra (1989) 'Phenomenology of small violations of Fermi and Bose statistics', *Physical Review* **D39**, p. 2035. Copyright (1989) by the American Physical Society.

horizontal boxes correspond to particles in symmetric states and the vertical boxes to particles in antisymmetric states. In paronic statistics of order 2, the paronic ground states for the first four elements of the second period of Mendeleev's table were predicted to be as indicated in Table 5.1. Thus, an atom with atomic number $Z$ in a representation with a Young pattern having one row with two boxes behaved chemically like an atom with atomic number $Z - 1$; the same atom in a representation with two rows and two boxes behaved like an atom with $Z - 2$, and so forth. This meant that paronic helium atoms behaved like hydrogen since they could accept another electron in their K-shell, while paronic beryllium was supposed to behave like an inert gas. Detailed calculations of paronic helium states were made by Drake[27] and, in 1995, an experiment was run[28] to search for paronic helium states as predicted by Greenberg and Mohapatra. As a two-fermion system, helium was the simplest system to test eventual violations of the exclusion principle: a metastable state such as $1s2s\,^1S_0$, where one electron is in the ground state and the other one in the lowest excited state, has a symmetric total wave function, in contrast with Pauli's principle. And Pauli-violating helium states could be revealed by eventual shifts in the spectrum, detected through modern laser spectroscopic techniques. In Deilamian *et al.*'s experiment, a helium atomic beam travelled through an interrogation region where it was irradiated by photons and it fluoresced. The total flux of metastable atoms was measured by a current induced on a stainless steel plate. A continuous scan over the interrogation region $2^3S_1 - 3^3P_1$ of the spectrum could test the predicted Pauli-violating spectral

---

[27] Drake (1989).  [28] Deilamian, Gillaspy, Kelleher (1995).

Table 5.1 *Normal and anomalous (Pauli-violating) ground states for some elements of the second period of Mendeleev's table.*

| Element | Normal states obeying Pauli exclusion principle | State violating Pauli principle |
|---|---|---|
| $^3$Li | $1s^22s^1$ | $1s^3$ ($\beta^2$) |
| $^4$Be | $1s^22s^2$ | $1s^32s^1$ ($\beta^2$) |
|  |  | $1s^4$ ($\beta^4$) |
| $^5$B | $1s^22s^22p^1$ | $1s^32s^2$ ($\beta^2$) |
|  |  | $1s^42s^1$ ($\beta^4$) |
| $^6$C | $1s^22s^22p^2$ | $1s^32s^22p^1$ ($\beta^2$) |
|  |  | $1s^42s^2$ ($\beta^4$) |
|  |  | $1s^32s^3$ ($\beta^4$) |

*Source:* Reprinted with permission from O. W. Greenberg and R. N. Mohapatra (1989) 'Phenomenology of small violations of Fermi and Bose statistics', *Physical Review* **D39**, p. 2034. Copyright (1989) by the American Physical Society.

lines (Fig. 5.4). Yet the digital scan over the interrogation region gave no indication of any Pauli-violating line.

Thus, the first line of attack against the exclusion principle (i.e. search for paronic ground states) did not give the expected results. What about the second line of attack (i.e. search for forbidden transitions)? In this case, Goldhaber and Scharff-Goldhaber's experiment in 1948 had already revealed a limit on the possibility of forbidden transitions in K-shell X-rays. But the first high-precision test came only in 1990 with Ramberg and Snow's experiment.[29] The significance of this test for the validity of the exclusion principle was stressed at the beginning of the article:

> In the past few years there have been several theoretical re-examinations of the question: 'Can the Pauli principle be violated by a tiny amount?'. Although no acceptable theoretical formulation has emerged from these investigations, the authors have stressed the importance of carrying out more rigorous experimental tests of the Pauli principle. L. B. Okun states that 'the absence of quantitative experimental tests of the Pauli principle in atoms is like a blank spot on the map of experimental physics'. Greenberg and Mohapatra have examined all past experimental data and concluded from these that the probability that a new electron added to an antisymmetric collection of $N$ electrons to form a mixed symmetry state rather than a totally antisymmetric state is $\leq 10^{-9}$. The purpose of this letter is to report an experiment in which this limit is reduced to $\leq 10^{-26}$.[30]

---

[29] Ramberg and Snow (1990).    [30] *Ibid.*, p. 438.

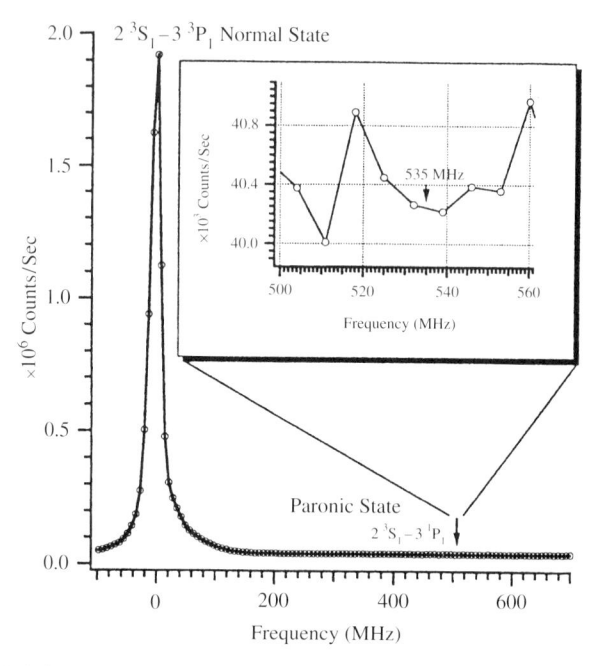

Fig. 5.4 Schematic diagram of Deilamian et al.'s experiment: fluorescence spectrum of helium scanned over the region of interest where paronic (Pauli-violating) lines were expected.

*Source:* Reprinted with permission from Deilamian, K., Gillaspy, J. D., and Kelleher, D. E. (1995) 'Search for small violations of the symmetrization postulate in an excited state of helium', *Physical Review Letters* **74**, p. 4789. Copyright (1995) by the American Physical Society.

Ramberg and Snow's experiment followed Goldhaber's original idea: given an electric current passing through a copper strip, K-shell X-rays would be emitted if one of the electrons were captured by a copper atom and cascaded down to the 1S state, despite the 1S state being already filled with two electrons. The eventual X-rays were expected to be shifted by about −0.6 keV from the 8.04 keV of the copper K-shell X-rays. Ramberg and Snow used Greenberg and Mohapatra's parametrization: they assumed that the probability of anomalous symmetric states was $1/2\beta^2$ and the probability of antisymmetric states was $1 - 1/2\beta^2$. They proceeded then to fix a limit on $\beta^2$ through the following experiment, run at Fermilab. An X-ray detector was situated above a thin strip of copper that was connected to a power supply, which could produce a current in the strip. The data set collected for two months had 142K triggers with the current at a steady 30 ampere for 24 hours alternating with current off for the same

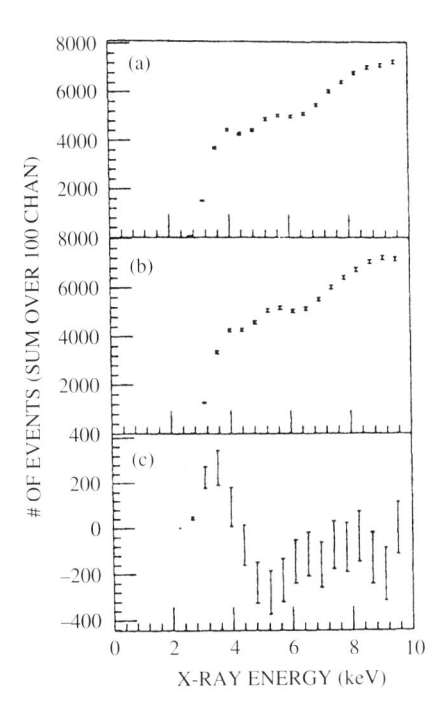

*5.3 Revoking or retaining the exclusion principle*     161

Fig. 5.5 Results of the Ramberg–Snow experiment. (a) Number of triggers summed over 100 ADC channels, plotted versus equivalent X-ray energy with current-on in copper strip below X-ray counter. (b) Same as (a) but with no current passing through an identical strip of copper. (c) Difference between (a) and (b) after normalization at the 9.5 keV point.

*Source:* Reprinted from E. Ramberg and G. Snow (1990), 'Experimental limit on a small violation of the Pauli principle', *Physics Letters* **B238**, p. 440. Copyright (1990), with permission from Elsevier.

period. To calculate the limit on $\beta^2$ from these data, all spectra were normalized to the number of triggers at 9.5 keV, and 100 channels of data – with current on and off, respectively – were summed over (Fig. 5.5). Ramberg and Snow found that the parameter $\beta^2$ was bounded by about $\leq 3.3 \times 10^{-26}$. They concluded:

> We have found that *there is no significant X-ray signal* above background coming from a piece of copper carrying current. We can interpret this as a limit on the strength of Pauli Exclusion Principle violating interactions between external electrons and copper atoms of magnitude less than $1/2\beta^2 = 1.7 \times 10^{-26}$.[31]

---
[31] *Ibid.*, p. 441. Emphasis added.

Thus, the second line of attack against Pauli's principle also did not yield the expected result. Neither Pauli-violating ground states nor transitions could be detected. These negative experimental results together with some theoretical problems of paronic quantum field theory (namely, the fact that many-particle states of this theory had negative squared norms) led Greenberg, a few years later, to substitute quons[32] for parons. Quons too are hypothetical particles with intermediate states $-1 < q < 1$, where $q = 1$ corresponds to Bose statistics and $q = -1$ to Fermi statistics. Their formalism, the quon algebra, interpolates between the Bose and Fermi algebras. Like parons, quons too would violate the exclusion principle only by a small amount. However, quon theory is affected by a no less serious problem: the observables of the theory violate locality, and consequently no relativistic quon field theory was possible.

The debate on the strict validity of the Pauli principle and the experimental limit on its violation is still on-going.[33] I shall not reconstruct the more recent developments of the parastatistics programme here, as it would lead me far astray from the purpose of this chapter. It suffices to note that the possibility of experimentally testing the exclusion principle in atomic physics emerged thanks to this research programme, whose philosophical implications for the validation of Pauli's principle I shall analyse in Section 5.4.1. Before turning to them let me first present the rival research programme, the one that reconciled negative evidence with the Pauli principle by introducing an additional degree of freedom for quarks.

### 5.3.2 Retaining the exclusion principle: coloured quarks and quantum chromodynamics

An alternative way out of the tension between quark theory and the exclusion principle was explored starting from an original suggestion by Greenberg. As mentioned in Section 5.3.1, Greenberg[34] proposed regarding quarks as parafermions of order 3. Shortly afterwards, this was reinterpreted as a new threefold degree of freedom for quarks, which later became known as 'colour'. Each quark was supposed to come with a colour that distinguished it from other 'equivalent' (same flavour) quarks. For instance, the three s quarks of the $\Omega^-$ particle were reinterpreted as one 'red', one 'green', and one 'blue'. The new degree of freedom was associated with a group of transformations, a double SU(3) symmetry, the so-called $SU(3)_c$ colour symmetry.

---

[32] Greenberg (1999).   [33] See Hilborn and Tino (2000).   [34] Greenberg (1964).

## 5.3 Revoking or retaining the exclusion principle

It was Y. Nambu and M. Y. Han[35] who originally introduced a three-triplet model of quarks with double SU(3) symmetry. The three triplets, which Han and Nambu did not yet refer to as coloured quarks,[36] could antisymmetrize the otherwise symmetric quark wave function. Indeed, the total wave function could now be rewritten as the product of a space wave function, a spin wave function, *and* a colour wave function so that the combined space and spin wave function could still be symmetric under interchange of any two quarks of the same flavour (as experimentally observed in baryon spectra), provided that the colour wave function was antisymmetric.[37] In this way, the exclusion principle could retain its strict validity for quarks.

For Han and Nambu the most important aspect of the three-triplet model was the possibility of restoring integral charges for quarks, a desirable feature absent in Gell-Mann's model and which other physicists[38] had tried to restore. But it was only in 1973, in a seminal paper of Bardeen, Fritzsch, and Gell-Mann[39] that the problem of antisymmetrizing the wave function was solved by tripling the number of fractionally charged quarks. It was in this paper that the term 'colour' was first introduced; the neutral pion decay and the hadron to muon production ratio in electron–positron collisions were indicated as possible evidence in support of the hypothesis of 'colour'. And, as we shall see in the following sections, decisive empirical support for 'coloured quarks' came from these two pieces of evidence, among many others.

Quantum chromodynamics (QCD) was born when the Yang–Mills gauge theory[40] was proposed as a candidate for the dynamics of eight gauge vector fields behaving as an octet in the $SU(3)_c$. They corresponded to eight quanta with electric charge 0, mass 0, and spin 1, called 'gluons' and carrying the 'chromodynamic force' binding quarks together in hadrons. Each gluon would carry a combination of a colour and an anticolour, for instance blue and antired. By emitting and absorbing gluons quarks would change colour (for instance, a quark u blue would become u red by emitting a gluon $G_{b\bar{r}}$).

---

[35] Han and Nambu (1965).
[36] Han and Nambu talked in terms of 'charm number'. However, the term 'charm' was later used to indicate a new quark flavour, as we shall see in Section 5.3.2.3.
[37] Each quark is associated with three independent colour wave functions represented by 'colour spinors' r, g, b. They are eigenfunctions of the 'colour operators', which are in turn represented by eight three-dimensional matrices. The total colour wave function for a baryon is then a linear superposition of six possible combinations, and it is antisymmetric, i.e. $(1/\sqrt{6})(rgb - grb + gbr - bgr + brg - rbg)$.
[38] Bacry *et al.* (1964). Triplets of integrally charged quarks were also proposed by Tavkhelidze (1965) and Miyamoto (1965). See Adler (2005).
[39] Bardeen, Fritzsch, and Gell-Mann (1973).   [40] Yang and Mills (1954).

Through the exchange of gluons the chromodynamic force is transmitted and quarks are kept bound in hadrons.

The coloured quark model could finally explain why only certain combinations of quarks and antiquarks were allowed out of all possible ones. As mentioned in Section 5.2, Gell-Mann identified baryons with quark triplets (qqq) and mesons with quark–antiquark pairs (q$\bar{\text{q}}$) without giving any theoretical justification for excluding other possible combinations such as qqqq or qq$\bar{\text{q}}$. The answer came from coloured quarks, more precisely from asymptotic freedom and colour confinement.

Like QED, QCD takes the form of a perturbation theory, whose coupling constant $\alpha_s$ has a peculiar property: when the momentum transfer $q^2$ increases asymptotically, the coupling constant $\alpha_s$ goes to zero. As a result, in the high-energy region, quarks behave *as if* they were free particles: they enjoy the so-called asymptotic freedom.[41] But at low energies such as those we find in nature, the coupling constant increases so that coloured quarks cannot exist as free particles:[42] they are confined inside hadrons. According to colour confinement, only colour singlet states, i.e. states with zero values for all colour charges, can exist as free particles, and these are precisely quark–antiquark pairs (mesons) and quark triplets (baryons).[43]

Given colour confinement, there is no hope of ever detecting free quarks in the lab. How could the auxiliary assumption of 'colour' for quarks become accredited? Was it just a useful auxiliary assumption suitably introduced to shelter the validity of Pauli's principle from prima facie negative evidence? Or was there any compelling reason for retaining the principle *in that specific way*, i.e. by introducing colour for quarks? In the

---

[41] Asymptotic freedom was theoretically predicted by Gross and Wilczek (1973), and Politzer (1973). Experimental evidence for it was found in the late 1970s with the detection of scaling violations (see next section). In 2004, Gross, Wilczek, and Politzer received the Nobel Prize for this discovery.

[42] Note that saying that coloured quarks do not exist as free particles does not mean that they cannot be knocked out of their bound states. In fact, they are typically knocked out of their bound states in experimental situations. But once knocked out, they recombine with other quarks to form new hadrons, rather than existing as free particles.

[43] More precisely, since $(1/\sqrt{3})(r\bar{r} + g\bar{g} + b\bar{b})$ is a colour singlet, q$\bar{\text{q}}$ entities (i.e. mesons) can exist as free particles. Similarly, $(1/\sqrt{6})(\text{rgb} - \text{grb} + \text{gbr} - \text{bgr} + \text{brg} - \text{rbg})$ is also a colour singlet state, so qqq entities (i.e. baryons) can exist as free particles. Other quark bound states whose total colour charge does not add up to zero do not exist as free particles. On the other hand, there are combinations such as qq$\bar{\text{q}}\bar{\text{q}}$ and qqqq$\bar{\text{q}}$ which are not forbidden by colour confinement. The debate on the alleged 'observations' of pentaquark bound states is ongoing and still not settled. Like quarks, gluons too do not exist as free particles because they are the carriers of net colours. Nonetheless gluon bound states whose total colour charge is zero (such as gg, gggg, etc.) can in principle exist: they are called glueballs and some experimental evidence has been found for them (see Close 1997).

following subsections, I shall briefly survey some of the main routes through which the auxiliary assumption of colour for quarks came to be accredited.

#### 5.3.2.1 Deep inelastic lepton–hadron scattering and scaling violations

Experimental evidence for coloured quarks came from some investigations that were designed and carried out quite independently of the theoreticians' work on coloured quarks. In the 1960s it was found that the inner structure of protons and neutrons was not uniform, but consisted of point-like entities whose nature and properties were investigated in the following ten years. The experimental technique to probe the inner structure of nucleons was deep inelastic lepton–hadron scattering, where a leptonic beam (electrons, muons, or neutrinos) is scattered off a hadronic target (e.g. protons or neutrons). Depending on the momentum transfer between the incident beam (say, electrons e), and the nucleon target (say, a proton p), the following two situations may occur: (1) at low momentum transfer, ep → ep (elastic scattering); (2) at high momentum transfer, ep → eX (deep inelastic scattering), i.e. the proton breaks up into many particles, where X is the set of hadrons allowed by conservation laws. Deep inelastic electron–proton and neutrino–proton scattering experiments offered the first evidence for the existence of inner constituents of nucleons. The experimental data were originally cast within the so-called parton model and only later reinterpreted in the light of the coloured quark model.

The coloured quark model was developed on mainly theoretical grounds, and independently of the deep inelastic scattering experiments that at more or less the same time were giving evidence for point-like entities inside nucleons, which were initially identified as 'partons'.[44] However, as soon as experiments showed that the spin and the electric charges of partons were identical to those of quarks, it seemed natural to identify partons with quarks. No wonder that, in the physics literature of the early 1970s, physicists often used the double name 'parton quark' and maintained an ambiguous terminology for their experimental findings.

Partons, however, could be identified with quarks only to a certain extent. According to the parton model, partons would not radiate gluons and exchange chromodynamic force: they would instead behave as free particles inside nucleons. The naïve parton model assumed that there were point-like partons $i$ inside the proton carrying a charge $e_i$. The partons

---

[44] Bjorken and Paschos (1969); Feynman (1972).

struck by the incident beam were supposed to move parallel to the parent proton (i.e. null transverse momentum). Each parton $i$ was supposed to carry a fraction $x$ of the proton's momentum $P$. And the probability that the struck parton $i$ carried a momentum fraction $x$ was given by the parton momentum distribution, which was predicted to be the same at all momentum transfers: i.e. it satisfied the so-called *Bjorken scaling* or *scale invariance*. Scaling behaviour in deep inelastic scattering was observed at SLAC in the late 1960s. The presence of point-like particles inside the proton was revealed precisely by this crucial mathematical property of the structure functions of being invariant – at a given value of the scaling variable $x$ – for all momentum transfers.

Thus, the dynamic properties of partons differed remarkably from those of coloured quarks: in this respect, partons and quarks could not be identified. And in 1979 the CDHS experiment[45] gave the verdict to coloured quarks[46] by detecting scaling violations: the violation of Bjorken scaling was the signal that gluons were emitted, and accordingly that coloured quarks (rather than partons) were the entities composing nucleons.[47] Quarks, originally introduced on mainly theoretical grounds, ended up stealing – so to speak – empirical support from partons, which were then discarded as potential constituents of nucleons. Yet there was a kernel of truth in the naïve parton model, as its relative success for almost a decade shows. One of the main achievements of QCD consisted in explaining why the parton model had been relatively successful: namely, at high-momentum regimes (like those of deep inelastic scattering), the QCD coupling constant

---

[45] See de Groot *et al.* (1979). CDHS is the acronym of the collaboration among CERN, Dortmund, Heidelberg, and Saclay.

[46] I have investigated this historical episode within the context of a criticism of Ian Hacking's experimental realism in Massimi (2004a).

[47] Scaling violations imply that the nucleon structure function $F_2$ does not remain invariant by varying the high-momentum transfer $Q^2$. Rather, the dependence of $F_2$ on $\log Q^2$ indicates that as $Q^2$ increases, the incident beam starts to probe better the inner structure of the nucleon. If the inner constituents were non-interacting partons, an increasing $Q^2$ would not resolve further structure, and Bjorken scaling would retain its validity. But if the inner constituents are coloured quarks exchanging gluons and producing an increasing sea of quark–antiquark pairs as the scaling variable $x$ approaches 0, then by increasing $Q^2$ we should be able to probe further structure inside the nucleon. In particular, we should be able to 'see' an increasing cloud of 'soft' quarks that have lost most of the nucleon's original momentum by radiating gluons. What we then expect to 'observe' is that at a fixed value of the scaling variable $x$, the structure function $F_2$ does not remain invariant by increasing $Q^2$. Rather, it either (1) decreases (at large $x$ values, i.e. $x \approx 1$) because the emitted gluon carries off a portion $x$ of the initial quark momentum $y$ so that the valence quark distribution shrinks towards smaller values of fractional momentum $x/y$ as $Q^2$ increases; or (2) the structure function increases (at small $x$ values, $x \approx 0$, where the sea of quark–antiquark pairs is dominant) because the sea distribution increases with increasing $Q^2$. This was exactly the kind of scaling violation detected in the CDHS experiment.

$\alpha_s$ tends to zero, and hence quarks behave *as if* they were free, non-interacting, parton-like entities because of asymptotic freedom.

### 5.3.2.2 Between gauge invariance and renormalizability: the neutral pion decay and the Adler–Bell–Jackiw anomaly

Empirical support for coloured quarks also came from another, more theoretical route. Following Yang and Mills' seminal work, gauge theory turned out to provide an encompassing theoretical framework for the weak and the strong interactions as well as the electromagnetic interaction. Yang and Mills' gauge theory was a field theory of strong interactions modelled upon QED. However, by contrast with QED where the photons, carriers of the electromagnetic interaction, are massless as required by such a long-range interaction, in the Yang–Mills gauge-theory Lagrangians the presence of massless terms constituted a stumbling-block for an understanding of the weak and strong interactions, whose short-range domain is incompatible with massless quanta.

In the case of the weak interaction, this problem was overcome thanks to spontaneous symmetry breaking, whose discovery paved the way to a unified gauge theory of the electroweak interaction, dating back to the pioneering work of Glashow,[48] and Salam and Ward,[49] and culminating in the Weinberg–Salam model in 1967. In this model, the intermediate vector bosons, carriers of the weak interaction, acquire mass through the so-called Higgs mechanism.[50] As 't Hooft proved,[51] gauge theories in which vector bosons acquire mass via the Higgs mechanism are renormalizable.

Renormalizability is a crucial feature of quantum field theory: it guarantees that the theory is well-behaved, i.e. that the terms in the perturbation expansion are non-divergent at high energies and to high orders in the coupling constant. The technique consists in introducing a finite number of (experimentally determined) parameters (including mass scales) that can cancel eventual divergences.

Gauge invariance, on the other hand, is an important symmetry classically associated with charge conservation. Quantum chromodynamics is a gauge theory of the strong interaction: the octet of gluons satisfies a local gauge invariance of quark colour, in analogy with the local gauge

---

[48] Glashow (1961). [49] Salam and Ward (1964).
[50] Higgs (1964a), (1964b). The massless Yang–Mills gauge bosons are supposed to acquire mass by interacting with the so-called Higgs field, a scalar field having a non-zero expectation value in its vacuum state, and not invariant under a gauge transformation (hence the spontaneously broken gauge invariance). The experimental detection of the Higgs boson represents one of the frontiers of high-energy physics.
[51] 't Hooft (1971).

invariance of quark flavour satisfied by the intermediate vector bosons of the electroweak theory. But gauge invariance can be spoiled by quantum effects: this is known as anomalous or quantum mechanical symmetry breaking.[52] As Adler[53] and, independently, Bell and Jackiw[54] discovered in 1968, quantum effects lead to broken gauge invariance for axial-vector currents, which are four-vector currents whose linear combinations with vector currents give electroweak currents.[55] As Adler recalls:

> the anomaly could not be eliminated without spoiling either gauge invariance or renormalizability ... my paper broke new ground by treating the anomaly neither as a baffling calculational result, nor as a field theoretic artifact to be eliminated by a suitable regularization scheme, but instead as a real physical effect (breaking of classical symmetries by the quantization process) with observable physical consequences ... These were the first indications that neutral pion decay provides empirical evidence for what we now call the 'colour' degree of freedom of the strong interactions.[56]

The neutral pion is a meson consisting of up (and down) quarks (or antiquarks); its two photon decay is described by a diagram with quarks running around the sides of a closed fermion loop (Fig. 5.6a), where the solid lines are quark fermion fields, and the curly lines are the two photons resulting from the decay $\pi^0 \to \gamma\gamma$. Axial-vector current conservation (or more precisely, partial conservation) relates this diagram to a second diagram in which the incoming $\pi^0$ is replaced by an axial-vector current A (Fig. 5.6b). Without going into technical details, it suffices here to say that this latter kind of closed fermion loop diagram leads to pathologies that are the core of the Adler–Bell–Jackiw anomaly. To satisfy axial-vector gauge invariance (i.e. current conservation or partial conservation) the $\pi^0 \to \gamma\gamma$ decay amplitude had to be vanishing.[57] But a non-vanishing result was experimentally found instead. The calculated decay amplitude, taking account of the anomaly, was in fact non-vanishing and proportional to the average of the electric charges of the u and d quarks involved:

---

[52] See Jackiw (1999).   [53] Adler (1969).   [54] Bell and Jackiw (1969).
[55] In the Fermi theory of $\beta$ decay mentioned in Chapter 4, weak interactions can be described as interactions of two currents (in analogy with electromagnetism) described by four-vector operators that, depending on their transformation properties under spatial reflections, could be vector ($V$), axial-vector ($A$), scalar ($S$), pseudoscalar ($P$), and tensor ($T$). With the discovery of parity violations in 1957, it was found that weak interactions couple through a left-handed mixture of vector $V$ and axial-vector $A$ currents.
[56] Adler (2004), pp. 5, 7.
[57] The relevant axial-vector current here is one that is not locally gauged in the electroweak theory. Anomaly cancellation between quarks and leptons eliminates the anomalies in diagrams involving locally gauged axial-vector currents, to which the intermediate vector bosons that mediate the weak interaction are coupled.

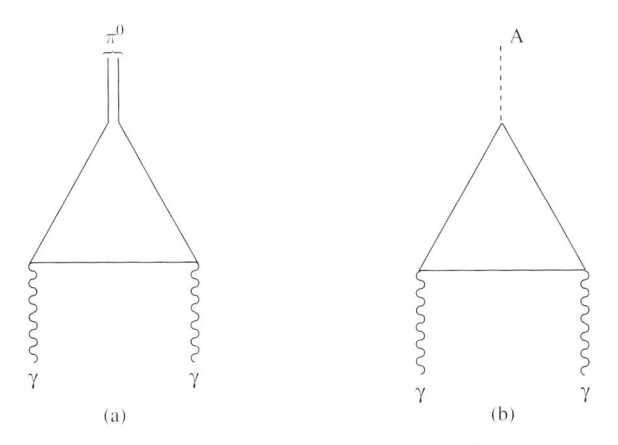

## 5.3 Revoking or retaining the exclusion principle

Fig. 5.6 (a) Closed fermion loop diagram for the decay $\pi^0 \to \gamma\gamma$. (b) Diagram related to the pion decay diagram by axial-vector current gauge invariance or partial conservation. This diagram has anomalous gauge-invariance properties.

*Source:* Drawn by and reproduced with kind permission of Stephen Adler.

$$\left(\frac{2}{3}\right)^2 - \left(-\frac{1}{3}\right)^2 = \frac{1}{3} \tag{5.2}$$

But this resulted in a decay rate that disagreed with the experimentally found value, which required the average of the electric charges to be equal to 1. Colour set everything right. By assuming that quarks come in three different colours, $1/3 \cdot 3 = 1$, the calculated decay amplitude turned out to agree with experimental data. Thus, via the Adler–Bell–Jackiw anomaly, the neutral pion decay could be used as a test for the charge structure of quarks (Gell-Mann's fractionally charged model versus Han and Nambu integrally charged model) as well as for colour.[58]

---

[58] It is important to clarify that the neutral pion decay offered independent empirical evidence for an extra factor of 3 for quarks. But this could be interpreted either as colour or as Green components of an order-3 parafermi field (see Greenberg 1993, p. 11). However, while colour can be readily incorporated in gauge theory, neither Green's formulation of parastatistics nor Messiah–Greenberg's could be gauged. Only in 1983, when QCD was already well established, could Greenberg and Macrae find a locally gauge-invariant version of parastatistics. So, in order to have a gauge theory of strong interaction, the 'colour' option was favoured (I thank Stephen Adler for illuminating comments on this point). Furthermore, the gauge theory of colour SU(3) was also required by the property of asymptotic freedom, and, as we saw in the previous section, empirical evidence for asymptotic freedom was found in the late 1970s with the discovery of scaling violations and the ability of QCD to explain why the parton model was successful in the high-energy region.

### 5.3.2.3 The ratio of cross-sections for hadron and muon production and a new quark flavour

Coloured quarks were experimentally accredited also through a third route. In electron–positron collisions, the energy released (the so-called centre-of-mass energy)[59] transforms into the masses of new particles, which appear as two oppositely directed jets via a two-step process involving the creation of a quark–antiquark pair first and its subsequent 'fragmentation' into the observed jets. Jets of muons and hadrons can be produced in electron–positron annihilations. And the ratio of the cross-sections[60] $\sigma$ for hadron and muon production,

$$R = \frac{\sigma(e^+e^- \to \text{hadrons})}{\sigma(e^+e^- \to \mu^+\mu^-)} \tag{5.3}$$

is proportional to the sum of the squared charges of the quarks produced in the electron–positron annihilations. Thus, a measurement of $R$ was expected to give important information about the number and nature of quarks.

At the beginning of the 1970s, the parameter $R$ was measured in the ADONE ring at Frascati (Italy), and later in the Cambridge Electron Accelerator (CEA) at Harvard. Since the known quarks at the time were only three (the u, d, and s quarks) with fractional charges 2/3, −1/3, and −1/3 respectively, the expected $R$ value was calculated as

$$R = (2/3)^2 + (-1/3)^2 + (-1/3)^2 = 2/3 \tag{5.4}$$

However, the experimental data showed that in the low-energy regime (between 1 and 3 GeV of the centre-of-mass energy) the ratio $R$ was bigger than expected. In particular, as Silvestrini[61] reported at the International Conference on High Energy Physics in September 1972, the measured value of $R$ was close to 2, suggesting that the number of events in which

---

[59] The centre-of-mass energy is the sum of the rest masses of the interacting particles plus their kinetic energies. Depending on the nature of the collision, all or just some portion of the centre-of-mass energy transforms into the masses of the final particles that can be calculated via the laws of momentum and energy conservation.

[60] In any scattering experiment, the rate of production of final particles can be measured by summing the contributions of the interactions of individual particles in the incident beam with individual particles in the target. This rate is proportional to the number of particles in the target hit by the beam, and to the rate per unit area at which the beam particles cross the target. The cross-section $\sigma$ of a reaction gives the dimensions of the area of the target surface crossed by the beam particles. And since this area is Lorentz invariant in the beam direction, the cross-section has the same value in the laboratory and centre-of-mass frame.

[61] Silvestrini (1972).

hadrons were produced was twice the number of muonic events. This discrepancy could be easily solved by introducing colour for quarks. If each quark flavour were associated with three possible colours, the ratio $R$ could be recalculated as

$$R = (2/3)^2 + (-1/3)^2 + (-1/3)^2 = 2/3 \times 3 = 2 \qquad (5.5)$$

in agreement with the experimental value. Thus, the introduction of the new degree of freedom for quarks was experimentally supported by data on $R$.

The parameter $R$ played also a major role in the following developments of QCD. Between 1972 and 1974, new experimental data on $R$ were gathered at the new electron–positron collider SPEAR at Stanford, whose results filled the gap between the earlier measurements at ADONE and CEA, and showed that the value of $R$ rose smoothly from 2 up to 6, by increasing the centre-of-mass energy $E_{cm}$ in the collisions from 2 to 5 GeV. In other words, in the high-energy regime (above 3 GeV), $R$ seemed to take up unexpected values revealing new phenomena. In the autumn of 1974, Richter and Goldhaber[62] observed an unexpectedly high peak in the scattering graphs, indicating that the number of hadronic events in electron–positron annihilations was not just double, but much higher than the number of muonic events, pointing to an unexpected value of $c$. 10/3 for $R$. The peak indicated the existence of a very narrow resonance[63] centred at 3.1 GeV with an unexpectedly long lifetime (almost 1000 times longer than one would expect given the heavy mass of 3.1 GeV). This narrow resonance, which Richter named $\psi$(psi), had already been spotted, a few days earlier, by another team of scientists led by Samuel Ting[64] at the Brookhaven synchrotron in Long Island (New York), working on the collision of protons against a beryllium target (p + Be → e$^+$ + e$^-$ + $x$). The peak was identified with a massive particle that Ting named J: to give both Ting and Richter their due, the particle became known as J/$\psi$.[65] Shortly after, the particle was identified with charmonium,[66] i.e. a bound state of a fourth quark flavour 'charm' and of its antiquark, whose existence had been predicted by Bjorken and Glashow.[67] In 1970, the Glashow–Iliopoulos–Maiani theory (GIM)[68] showed

---

[62] See Augustin, Richter *et al.* (1974).
[63] The term 'resonance' goes back to Fermi's discovery of the resonance $\Delta^{++}$. Resonances have not only well-defined masses, but also well-defined quantum numbers such as isospin, and strangeness. Despite the name, they are particles to all intents and purposes.
[64] See Aubert, Ting *et al.* (1974).
[65] For a reconstruction of the discovery of the particle and following developments between 1974 and 1976, see Ritcher (1977).
[66] Appelquist and Politzer (1975). [67] Bjorken and Glashow (1964).
[68] Glashow, Iliopoulos, and Maiani (1970).

that introducing charm led to suppression of various strangeness-changing neutral current effects. In this respect, GIM amended Cabibbo's theory, which as we saw in Section 5.2 allowed the existence of strangeness-changing weak neutral currents that, however, were observed to occur very rarely in nature. Thus, the discovery in 1974 of a value of the order of 10/3 for $R$ in the high-energy regime disclosed the existence of another quark flavour (charm), and indirectly offered further evidential warrant for colour. The new quark flavour c has electric charge 2/3 like the u quark, so the expected $R$ value including the quark c contribution becomes:

$$R = (2/3)^2 + (2/3)^2 + (-1/3)^2 + (-1/3)^2 = 10/9 \qquad (5.6)$$

which multiplied by a colour factor 3 becomes 10/3, as experimentally observed in Richter's scattering graph.

## 5.4 The Duhem–Quine thesis: epistemological holism and the validation of the exclusion principle

The two rival research programmes of parastatistics and quantum chromodynamics stemmed from the difficulty of reconciling – within the quark theory – the exclusion principle in its enlarged nomological scope given by the spin–statistics theorem on the one hand, with the negative evidence of the $\Omega^-$ particle and the baryon spectra on the other hand. Both programmes could accommodate these pieces of negative evidence. Should we then reconcile recalcitrant evidence by revoking the strict validity of the exclusion principle for quarks, or rather by introducing the auxiliary assumption of colour for quarks? The Duhem–Quine thesis here comes to the fore.

As is well known, the French philosopher and physicist Pierre Duhem and the American philosopher Willard van Orman Quine put forward a very similar thesis that, to give both their due, has since become known in the philosophical literature as the Duhem–Quine thesis.[69] The thesis says that, given the holistic character of scientific knowledge, where the main theoretical assumptions are always conjoined with auxiliary ones to make predictions, it is never possible – in the case of negative evidence – to put the blame for it on the main theoretical assumptions any more than on the auxiliary ones. In other words, in the case of negative evidence, the choice

---

[69] Duhem (1906); Quine (1951).

## 5.4 The Duhem–Quine thesis and the validation of the exclusion principle

between rejecting the core of the theory (including scientific principles) or retaining it by introducing suitable auxiliary assumptions cannot be rationally settled, and remains at the discretion of scientists' 'good sense' in Pierre Duhem's words. The Duhem–Quine thesis challenges the rationality of theory-choice, or, to put it in a slightly different way, the rationality of retaining threatened theories (by introducing auxiliary assumptions) when the reasons for retaining are as strong as the reasons for rejecting them. Epistemological holism underlies the Duhem–Quine thesis. As mentioned in Section 1.3, according to Quine's epistemological holism, scientific knowledge (from the central areas of logic and arithmetic up to the peripheral areas occupied by more empirically oriented disciplines, such as chemistry or biology) is a vast web of interconnected beliefs that impinges on experience only along the edges.[70] Epistemological holism has traditionally been regarded as a royal road to underdetermination of theory by evidence because it sheds light on the ambiguity of potentially falsifying instances. As positive instances are unable to confirm one hypothesis over another, both implying the same empirical consequences, similarly negative instances seem unable to falsify one hypothesis over another.[71]

In the specific historical episode reconstructed in this chapter, a supporter of Quinean holism might then argue that the enduring nomological validity of the exclusion principle in the face of recalcitrant evidence can ultimately be traced back to its being a well-accredited principle of atomic and nuclear physics and to its occupying a fairly stable, and hence less prone to revision, position in the web of knowledge. Coloured quarks could be just the expression of scientific conservatism, of a methodological decision to retain the Pauli principle 'come what may' in high-energy physics. Andy Pickering,[72] for instance, has claimed that the reasons fo the increasing success of QCD with respect to parastatistics must be looked for in the scientists' preference for familiar theories over new and obscure formalisms (such as parastatistics) and in the opportunities of research in a particular social context. Steven French[73] has responded to

---

[70] Quine (1951), reprinted in Quine (1953), pp. 20–46.
[71] Laudan (1990) refers to the first as 'Humean underdetermination' typically affecting confirmation theories such as hypothetico-deductivism, and to the second as 'Quinean underdetermination' affecting ampliative inferences. I have discussed the specific threat that underdetermination poses in the context of hypothetico-deductivism in Massimi (2004b), where I defend demonstrative induction as a promising non-ampliative inference to forestall underdetermination. Here I want to focus instead on Quinean underdetermination (Section 5.4.2).
[72] Pickering (1984), chapter 7.   [73] French (1995).

174    5 *From the eightfold way to quantum chromodynamics*

Pickering that quantum chromodynamics has undeniable theoretical merits over parastatistics, namely its being gauge invariant and unexpectedly fruitful in generating new lines of research.

In what follows, I do not intend to take up the cudgels for QCD. I shall not claim that QCD has been experimentally proved and parastatistics refuted, or that the two programmes did not both have a rationale. Rather, I want to address the philosophical challenge posed by the Duhem–Quine thesis in this specific historical episode. The upshot is to show that epistemological holism is not necessarily bad news for the rationality of theory-choice. In particular, I am going to argue that a holistic conception of scientific knowledge, one that regards scientific theories as clusters of intertwined main theoretical assumptions and auxiliary ones, can on the contrary be epistemically fruitful: it can account for the way in which the nomological validity of the exclusion principle came to be strengthened, and hence it can ground the rationality of retaining the principle in the face of negative evidence.

The story told in this chapter can accordingly be seen from a different perspective. It is not only the story of two rival research programmes prompted by some negative evidence about the exclusion principle in the quark theory. It is also the story of how – thanks to the availability of these two rival programmes – the exclusion principle came to be nomically strengthened and experimentally validated. This process of validation was indeed a two-way street. On the one side, the availability of the parastatistics programme made it possible to predict possible violations of the principle in the context of atomic physics (e.g. paronic helium and copper). So the negative evidence for parons found in the 1990s offered evidential warrant for the strict validity of the principle. On the other side, the introduction of the auxiliary assumption of colour for quarks so as to reconcile the principle with negative evidence, turned out to be supported by a wide-ranging array of empirical evidence such as (i) evidence about scaling violations; (ii) evidence about the hadron-to-muon production ratio $R$; (iii) evidence about the neutral pion decay rate as predicted by the Adler–Bell–Jackiw anomaly; and (iv) more direct evidence such as hadron-jet and glue-jet events,[74] and the phenomenological models for them, which turned out to be dependent on QCD.[75] For reasons of space,

---

[74] See Hofmann (1981).
[75] Peter Galison (1997), pp. 648–61, analyses three phenomenological models for the hadronization processes as mediators between QCD and the plethora of experimental data: (1) the Feynman–Field fragmentation model, which did not draw on QCD; (2) Andersson's string model, more prone to accept QCD considerations and yet still more

## 5.4 The Duhem–Quine thesis and the validation of the exclusion principle

I cannot present here this last type of experimental practice or the many other empirical successes of QCD in the past twenty-five years.

Epistemological holism need not be the threat to the rationality of theory-choice that it has typically appeared to be. On the contrary, it can be good news for retaining Pauli's principle in the face of recalcitrant evidence, since (i) it allows pieces of negative evidence within the parastatistics programme to play a positive validating role for Pauli's principle, and (ii) it can convey and channel a wide-ranging array of experimental evidence to support the new auxiliary assumption of colour. I clarify the first point in the next subsection, and the second in the following one.

### 5.4.1 The validating role of negative evidence

From the point of view of epistemological holism, the exclusion principle can be regarded as trapped between two opposite poles. On the one hand, it is amenable to experimental revision, as any other part of a scientific theory; on the other hand, it is a fairly central element of the new quantum theory framework broadly construed and as such it is less prone to be modified or refuted in case of negative evidence. Nevertheless, the story told in this chapter clearly illustrates that the enduring nomological validity of Pauli's principle cannot be traced back to its being a presumably well-entrenched element of the new framework, *pace* any scientific conservatism. Rather, it must be traced back to its passing a series of experimental tests that have progressively reduced the limit for possible violations. Thanks to parastatistics, experimental tests of possible violations of the exclusion principle could finally be devised and run. Their negative results are what in the end underpins the enduring nomological validity of Pauli's principle. Let me clarify how negative evidence could play such a positive validating role for the exclusion principle.

As we have seen in Chapter 4, the exclusion principle passed from the status of a phenomenological rule to that of a well-accredited scientific principle of the new quantum mechanics framework after its reformulation as a veto to symmetrizing the wave function for an assembly of indistinguishable half-integral spin particles, or (in quantum field theory) as a veto to commuting Dirac field operators. Reformulated in this way, Pauli's rule came to play a crucial systematizing role in the new theoretical framework, by unifying more and more areas within its prescriptive domain.

experimentally oriented; (3) the Field–Wolfram cluster model, which on the contrary resorted significantly to QCD and beat the other two models when confronted with data about depletion in three-jet events.

176    5 *From the eightfold way to quantum chromodynamics*

Table 5.2

|  | $\|a^r\rangle$ | $\|a^s\rangle$ |
|---|---|---|
| Case 1 | 1, 2 |  |
| Case 2 (S) | 1 | 2 |
| Case 3 (A) | 1 | 2 |
| Case 4 |  | 1, 2 |

S: symmetric state vector
A: antisymmetric state vector

In this chapter, we have further seen how the veto expressed by the exclusion principle could be regarded as following from a more general prescription expressed by the indistinguishability postulate. As discussed in Section 5.3.1, this postulate allows higher symmetry types (para-Fermi and para-Bose) as well as the usual symmetric and antisymmetric ones. It is the mathematical availability of higher symmetry types, licensed by the indistinguishability postulate but usually ruled out by the symmetrization postulate, that prompted the development of parastatistics in the 1960s. The indistinguishability postulate expresses an important principle of invariance: permutation invariance, i.e. invariance under the permutation group, which underlies the statistical behaviour of an ensemble of indistinguishable particles.

To appreciate how permutation invariance affects the statistical behaviour of quantum particles, consider a system of only two indistinguishable particles (say, two photons or two electrons). Let us call the two particles 1 and 2, and let us denote the two quantum states they can be in as $|a^r\rangle$ and $|a^s\rangle$. There are four possible cases as shown in Table 5.2. Let us write the four cases more precisely in mathematical terms as follows

case 1: $|a_1^r\rangle \otimes |a_2^r\rangle$
case 2: $1/\sqrt{2}(|a_1^r\rangle \otimes |a_2^s\rangle + |a_1^s\rangle \otimes |a_2^r\rangle)$
case 3: $1/\sqrt{2}(|a_1^r\rangle \otimes |a_2^s\rangle - |a_1^s\rangle \otimes |a_2^r\rangle)$
case 4: $|a_1^s\rangle \otimes |a_2^s\rangle$

where the composite system of the two particles is given by the tensor product of their respective state vectors if the particles are in the same state $r$ or $s$ (cases 1 and 4 respectively); by the symmetric state vector (case 2) or by the antisymmetric state vector (case 3), if the particles are in different states $r$ or $s$.

Classical Maxwell–Boltzmann statistics (MB) would assign equal probabilities to the four cases as shown in Table 5.3. The situation changes drastically if we move to quantum statistics. Bose–Einstein statistics (BE)

## 5.4 The Duhem–Quine thesis and the validation of the exclusion principle

Table 5.3

|        | $|a^r\rangle$ | $|a^s\rangle$ | MB |
|---|---|---|---|
| Case 1     | 1, 2 |      | 1/4 |
| Case 2 (S) | 1    | 2    | 1/4 |
| Case 3 (A) | 1    | 2    | 1/4 |
| Case 4     |      | 1, 2 | 1/4 |

Table 5.4

|        | $|a^r\rangle$ | $|a^s\rangle$ | MB | BE |
|---|---|---|---|---|
| Case 1     | 1, 2 |      | 1/4 | 1/3 |
| Case 2 (S) | 1    | 2    | 1/4 | 1/3 |
| Case 3 (A) | 1    | 2    | 1/4 | 0 |
| Case 4     |      | 1, 2 | 1/4 | 1/3 |

Table 5.5

|        | $|a^r\rangle$ | $|a^s\rangle$ | MB | BE | FD |
|---|---|---|---|---|---|
| Case 1     | 1, 2 |      | 1/4 | 1/3 | 0 |
| Case 2 (S) | 1    | 2    | 1/4 | 1/3 | 0 |
| Case 3 (A) | 1    | 2    | 1/4 | 0   | 1 |
| Case 4     |      | 1, 2 | 1/4 | 1/3 | 0 |

gives an equal probability of 1/3 to cases 1, 2, and 4 (Table 5.4), because case 3 (antisymmetric state vector) does not apply. Similarly, Fermi–Dirac statistics introduces a non-classical assumption of equiprobability (Table 5.5). In this case, a probability of 1 is assigned to case 3, because case 2 (symmetric state vector) does not apply. Nor, on the other hand, do cases 1 and 4 apply either given Pauli's principle. So far, we have only considered two indistinguishable particles. But as soon as we consider three or more particles, we find that there are subspaces of the Hilbert space ('generalized rays') that are invariant under all permutations: Greenberg and Messiah took these subspaces as encoding higher (para-Fermi and para-Bose) symmetry types as we saw in Section 5.3.1.

Thus, permutation invariance is itself playing a systematizing role in the quantum mechanics framework in the sense of making the already known quantum statistics (Fermi–Dirac and Bose–Einstein) follow from a

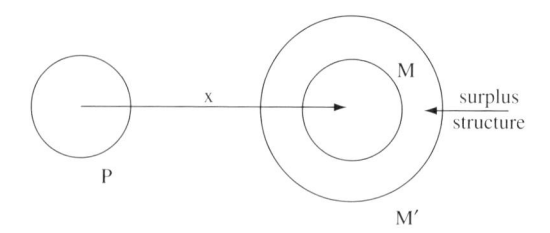

Fig. 5.7 'Surplus structure' in the representation of the physical structure P in the larger mathematical structure M'.

*Source*: Reprinted with permission from M. Redhead (2002) 'The interpretation of gauge symmetry', in H. Lyre, M. Kuhlmann, and A. Wayne (eds.) *Ontological Aspects of Quantum Field Theory* (Singapore: World Scientific). Copyright (2002) by World Scientific.

more general group-theoretical prescription that also discloses new symmetry types.[76] This is an example of 'surplus structure', to use Redhead's terminology:[77] a physical structure P (e.g. the quantum statistical behaviour of an ensemble of indistinguishable particles) is not represented by the mathematical structure M (e.g. Fermi–Dirac and Bose–Einstein statistics) with a one–one structure-preserving map between P and M. Rather, P is represented by a larger mathematical structure M' (e.g. permutation invariance), hence a surplus structure M' – M (e.g. generalized rays) in the representation of P by means of M' (Fig. 5.7). One interesting question concerning surplus structure is whether or not it can be invested with physical reality; in this case, whether the surplus structure (generalized rays) implied by permutation invariance possesses any ontological significance. The on-going experimental search for paraparticles, parons, and more recently quons seems to suggest a positive answer. I am not going to address this issue here. I want rather to draw attention to a different issue: namely, how the specific

---

[76] Giving a structuralist twist, one may claim that it is ultimately permutation invariance that binds the 'web of relations', as Steven French and Dean Rickles (2003), p. 233 have recently suggested: 'Permutation invariance can be seen as embodying a form of structuralist representation of "broader" kinds, namely bosons, fermions, parabosons, parafermions, and so on. In other words, the status of permutation invariance, from this perspective, is that of one of the fundamental symmetry principles which effectively binds the "web of relations" constituting the structure of the world into these broad kinds.' I find this structuralist reading of permutation invariance compatible and even congenial to the characterization I have given of the exclusion principle as playing a systematizing role in the new quantum theory framework broadly construed (from non-relativistic quantum mechanics to relativistic quantum field theory). Permutation invariance itself can be regarded as playing a systematizing role – albeit at a higher level – hence Pauli's veto follows.

[77] Redhead (1975), and Redhead (2003), p. 128–9.

## 5.4 The Duhem–Quine thesis and the validation of the exclusion principle

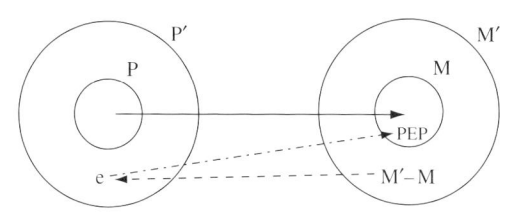

Fig. 5.8 'Surplus structure' playing a validating role in the case of Pauli's exclusion principle. The larger mathematical structure M′ disclosed by permutation invariance creates surplus structure M′– M (generalized rays), which entails (dashed line) empirical consequences e (paronic/Pauli-violating states) in a larger physical structure P′. The negative evidence found for parons in the 1990s confirmed the strict validity of Pauli's exclusion principle (PEP) – dotted line from e to PEP.

systematizing role of permutation invariance, which gives rise to 'surplus structure', has the positive effect of validating the exclusion principle.

More precisely, the surplus mathematical structure M′ – M (generalized rays) entails empirical consequences e (e.g. paronic/Pauli-violating helium and copper) in a broader physical structure P′ obtained by revoking a main theoretical assumption of M (e.g. Pauli's exclusion principle, PEP) (Fig. 5.8). And the negative experimental results for parons discussed in Section 5.3.1.1 confirmed the strict validity of Pauli's principle. In this way, central assumptions of the mathematical structure M (i.e. exclusion principle), which tend not to be subject to direct experimental testing, become in fact testable, as is to be expected in a truly dynamical conception of scientific knowledge. Experimental tests of a scientific principle may not be available in the original context, and may become feasible only at a later stage and within a broader framework. Only by questioning the strict validity of the exclusion principle in high-energy physics was it possible to devise experimental tests that confirmed its strict validity in atomic physics, namely in the very same domain in which the principle was originally introduced as an 'extremely natural' rule.

### 5.4.2 Quinean underdetermination and the rationality of retaining a threatened principle

So far I have been claiming that epistemological holism is not necessarily bad news for the rationality of theory-choice, and, on the contrary, it can be fruitful for validating a prima facie threatened principle. As I said, the validation process is a two-way street. In the previous section I have shown how a broader mathematical structure M′ disclosed by permutation invariance entailed empirical consequences – obtained by revoking Pauli's principle – in a broader physical structure $P'$ so that negative evidence for parons played a validating role for Pauli's principle. In this section I clarify the reasons why the introduction of the auxiliary assumption of colour was not a stratagem to rescue a well-entrenched principle. Or, to put it another way, I clarify why the enduring nomological validity of the exclusion principle, from atomic physics to nuclear physics to high-energy physics, cannot be reduced to its being a well-accredited theoretical assumption that some form of scientific conservatism would dissuade us from dismissing.

As mentioned in Section 5.4, epistemological holism has typically been regarded as a royal road to underdetermination. Given the holistic nature of scientific knowledge in facing the 'tribunal of experience', there seems to be enough space for manoeuvring and adjusting the web of beliefs so as to reconcile it with negative evidence. Larry Laudan[78] has given a penetrating analysis of 'Quinean underdetermination' (QUD), as the thesis that

> QUD: Any theory can be reconciled with any recalcitrant evidence by making suitable adjustments in our other assumptions about nature

hence two possible readings. On a weak *compatibilist* reading, the expression 'can be reconciled with' is synonymous with 'can be made logically compatible with the formerly recalcitrant evidence' by suitably stripping out any assumption that can be responsible for the recalcitrant evidence. On a strong *entailment* reading, it is synonymous with 'can be made to function significantly in a complex that entails' the formerly recalcitrant evidence by replacing a set of auxiliary assumptions with another that allows such a derivation. Laudan then concludes that neither logical *compatibility* with the evidence nor logical *derivability* of the evidence is sufficient to establish the rational acceptability of the theory so amended, because they both fail to establish either an explanatory relation or a relation of empirical support.

Building on Laudan's criticism, I want to argue against Quinean underdetermination that the rationality of retaining a threatened principle, such as Pauli's, in the face of recalcitrant evidence is ultimately grounded on the explanatory relation and on the relation of empirical support that the principle, suitably amended with the auxiliary assumption of colour, satisfies. Epistemological holism can in fact help us spell out the explanatory relation and the relation of empirical support that Pauli's principle, so amended, enjoys.

---

[78] Laudan (1990).

## 5.4 The Duhem–Quine thesis and the validation of the exclusion principle 181

Quine's web of beliefs is indeed a double-edged sword: it allows amending and adjusting assumptions, while it still has to face the 'tribunal of experience' along the edges. And the tribunal of experience supposedly does more work than just offering pieces of potentially falsifying evidence: presumably, it feeds the web with as many as possible pieces of evidence, which can in turn be used to assess the credentials of a threatened principle amended by introducing a new auxiliary assumption. In other words, epistemological holism need be no more pernicious than fruitful: it can help us to spell out the degree of empirical support that the web gives to a threatened principle once suitably amended. Thus, the rationality of retaining the exclusion principle *in that specific way* (i.e. by introducing colour for quarks) should be looked for in how QCD fared on the score of multifarious evidence and in the degree of empirical support that in so doing accrued to QCD.

As we have seen in Section 5.3.2, empirical support accrued to QCD from a wide-ranging array of evidence that passed through various channels – so to speak – from the *inside out* as well as from the *outside in*. The channel 'from the inside out' involves inter-theoretic links that make a theory amenable to being empirically supported from independent empirical evidence for other theories, to which the theory may be linked. On the other hand, the 'outside in' channel includes the vast class of experimental, more phenomenologically oriented investigations that may be carried out quite independently of the theoreticians' work. For instance, one of the 'inside out' channels of empirical support for the auxiliary assumption of colour passed through the renormalization of the electroweak theory. As discussed in Section 5.3.2.2, gauge theory provided the framework for quark–lepton interactions in the Weinberg–Salam electroweak theory as well as in QCD. The discovery in 1969 that quantum effects broke gauge invariance for axial-vector currents (the Adler–Bell–Jackiw anomaly) suggested that the renormalizability of the electroweak theory was spoiled. By introducing an extra factor 3 for the quarks, the anomaly could cancel between quarks and leptons in their locally gauged currents, as well as giving the correct neutral pion decay rate. Thus, these experiments offered independent evidence for the auxiliary assumption of colour insofar as this was required for the absence of anomalies in locally gauged currents, and hence for the renormalization of the electroweak theory. Via the inter-theoretic link provided by the gauge-theory framework, evidence for the renormalizability of the electroweak theory could give evidential warrant to QCD.

Another channel of empirical support 'from the inside out' was the prediction of charmed particles by the Glashow–Iliopoulos–Maiani theory

(GIM). As mentioned in Section 5.3.2.3, the detection of the J/psi in 1974, and of many other charmonia in the following years, confirmed GIM, and with it the existence of a fourth quark flavour (charm) that itself turned out to be available in three colours. Evidence for strangeness-changing weak neutral currents provided QCD with further empirical support.

Shifting to the channels 'from the outside in', substantial evidence for colour came from deep inelastic lepton–hadron scattering, originally cast in the parton model. As discussed in Section 5.3.2.1, these experiments were carried out quite independently of the parallel development of QCD, and in a period when coloured quarks were still only a tentative hypothesis on mainly theoretical grounds. It took almost ten years to arrive at a final identification of partons with quarks, or better to interpret the experimental evidence accumulated in that span as evidence for coloured quarks rather than for partons. It was the detection of scaling violations that finally discriminated between partons and quarks and shed light on the quantum chromodynamical properties of the hadron's constituents. At the same time, scaling violations could also explain why the parton model had been so successful for almost ten years: in the high-energy region, coloured quarks behave *as if* they are partons because of asymptotic freedom.

To sum up, the rationality of retaining the exclusion principle *in that specific way* (i.e. by introducing colour for quarks as an auxiliary assumption) is ultimately rooted in the way this auxiliary assumption has fared on the score of (1) being itself empirically supported by independent empirical evidence, and (2) increasing the explanatory scope of the quark theory by accounting for old evidence previously cast in the parton model, and predicting novel phenomena – scaling violations and charmonia among many others.

Pauli's principle amended with the auxiliary assumption of colour for quarks then satisfied both an explanatory relation and a relation of empirical support. Only under these conditions is it rational to retain a threatened principle, *pace* Quinean underdetermination. The rationality of retaining Pauli's principle in the quark theory can be defended in a way that does not make any concession to scientific conservatism. This rationality resides in the strategic intertwining of the principle with many other theoretical constraints as well as with a huge array of experimental evidence: this very same intertwining, which has traditionally appeared as opening the door to Quinean underdetermination, may in fact also do the opposite job of strengthening the principle in the face of possible recalcitrant evidence.

The exclusion principle is an important knot of a scientific web woven from threads of experimental evidence and theoretical constraints. To use

## 5.4 The Duhem–Quine thesis and the validation of the exclusion principle

a well-known metaphor of Wittgenstein, as the strength of a rope does not derive from one single thread running along the whole rope but from a multiplicity of small threads intertwined; similarly, the enduring nomological validity of the exclusion principle does not depend on its being a well-entrenched knot, but rather on the several (experimental and theoretical) threads it has successfully interwoven, and in the potentialities that in so doing it has disclosed.

# Conclusion

The aim of this book was to investigate the rationale for Pauli's principle, and the conditions under which we are justified in regarding a phenomenological and contingent rule as an important scientific principle. To this purpose, I urged a Kantian perspective. As a conclusion, I want to foreshadow how this perspective relates to contemporary discussions about images of science, and how it bears upon the on-going debate on scientific realism in philosophy of science. I shall barely scratch the surface of this complex topic and its vast literature. The best I can do in these concluding remarks is to suggest a philosophical position that I think deserves to be further explored. What follows should then be read more as an outline for future research, than as a conclusion.

I began this book by reconstructing the origins of the exclusion principle as a phenomenological rule arising from the spectroscopic research of the 1920s. I argued that the rule was derived – in conjunction with the concept of the electron's *Zweideutigkeit* – from spectroscopic phenomena with the help of some theoretical assumptions, in the period of revolutionary transition from the old to the new quantum theory. I defended the prospective intelligibility of this revolutionary transition in the light of the piecemeal process of transformation of the old quantum theory, from which brand new scientific concepts and nomic generalizations followed. The shaky theoretical grounds and the mainly phenomenological basis on which Pauli's exclusion rule was introduced could not justify its current nomological status as an important scientific principle. As we saw in Chapter 4, the process of accreditation of Pauli's rule passed through its reformulation in terms of Fermi–Dirac statistics in 1926. In this way, Pauli's principle was built into the new quantum theory framework from the ground up, wherein new laws were derived, such as the spin–statistics theorem, which in turn enlarged the scope of applicability of the principle to any half-integral spin particle.

I urged a Kantian perspective to interpret this nomological shift as grounded on the role that Pauli's rule came to play in the systematization of the new quantum theory framework. The rule – reformulated as a prescription to antisymmetrize the wave function – could accomplish the Kantian regulative task of systematizing the increasing body of quantum mechanical knowledge, and in so doing expanding the class of phenomena that could be captured. As we saw in Chapter 1, from a Kantian viewpoint the goal of systematization should not be understood just as an additional desideratum, but rather as the *conditio sine qua non* of identifying empirical regularities as lawlike and conferring necessity and nomological strength upon them. In this respect, the systematizing role that Pauli's principle accomplished in the new quantum theory can be regarded as fulfilling a Kantian regulative task. This Kantian perspective distances itself from the strong realist view that considers systematic unity as mirroring the true structure of nature as well as from the anti-realist counterclaim that regards it just as an additional pragmatic desideratum.

The final step consisted in showing that the systematization achieved by Pauli's principle within the new framework was not a mere 'projected unity', notwithstanding the regulative nature. To this purpose, I considered the process of validation of the principle that – as we saw in Chapter 5 – passed through both negative evidence for Pauli-violating states predicted by parastatistics, and positive evidence for Pauli-obeying coloured quarks predicted by quantum chromodynamics.

The overall picture of the origin and validation of the exclusion principle proposed starts from phenomena and to them leads back again, while entrusting the principle with the regulative task of conferring an inner systematization on the body of quantum mechanical knowledge, which ultimately impinges on these phenomena. This picture owes an obvious debt to Ernst Cassirer's view of scientific knowledge as a three-layer architectonic consisting of results of measurement, laws, and principles, and as such it is germane to a shift of focus from scientific entities to laws. Scientific entities no longer constitute the starting point from which laws of nature and scientific principles are read off. Rather, the other way around: phenomena and principles become the two complementary poles of a system of knowledge, within which only epistemic access to scientific entities is possible. Or, to echo Cassirer, 'the extent of the dominance of these laws marks the extent of our objective knowledge. Objectivity or objective reality, is attained only because and insofar as there is conformity to law – not vice versa ... Apart from this reality there exists for us no other objective reality to be investigated or sought after.'[1] Let me briefly clarify this point and its implication for the ongoing realist/antirealist debate in philosophy of science.

The debate on the truth of scientific theories is orthogonal to the debate on objectivity. From a traditional realist position, truth is regarded as correspondence with an objective, i.e. mind-independent reality. On the other hand, from an antirealist perspective, such as Bas van Fraassen's constructive empiricism, scientific theories do not aim at truth but only at saving phenomena, and the quest for objectivity is perceived as an erroneous tendency of scientific inquiry – as 'objectifying inquiry' – to reify its products, to give them the status of objects.[2]

One may wonder whether there is any way out from this realist/antirealist impasse, a third intermediate position, according to which objectivity is neither equated with objects already given independently of our scientific theorizing, nor reduced to a chimera of a science that tends to reify its products. I think that a Kantian perspective provides us with a welcome *via media*, equally distant from metaphysical realism as well as from antirealism. It licenses a mild realist position, according to which the boundaries of our scientific system – with phenomena as empirical basis and scientific principles playing an inner, systematizing role – define and fix the boundaries of objective reality itself. Ultimately, it is systematicity that underpins truth: there is no objective reality, and hence no truth, to be investigated and sought after over and above the boundaries fixed by a scientific system so defined.

A Kantian form of realism is not a novelty in philosophy of science. Hilary Putnam's internal realism[3] is the paradigmatic example.[4] On this view, the quest for truth is redefined as the quest for objectivity intended as

---

[1] Cassirer (1936). English translation (1956), p. 135.
[2] On 'objectifying inquiry', see van Fraassen (2002) pp. 160–4. The notion of 'objectifying inquiry' is inspired by Edmund Husserl's *Crisis of European Sciences*, upon van Fraassen's direct admission (*ibid.*, Footnote 3 of Lecture 5, p. 251). Nor are phenomenological considerations of this type new to van Fraassen. Indeed, they were already in *The Scientific Image* (1980), pp. 80–2. Antirealism (i.e. 'bracketing' the ontological implications of the scientific world-picture, and suspending any epistemic commitment) is then regarded as a sensible option to resist reification.
[3] 'On the [externalist] perspective, the world consists of some fixed totality of mind-independent objects. There is exactly one true and complete description of "the way the world is". Truth involves some sort of correspondence relation between words or thought-signs and external things. I shall call this perspective the externalist, because its favourite point of view is a God's eye point of view. The perspective I shall defend has no unambiguous name. It is a late arrival in the history of philosophy ... I shall refer to it as

'objectivity for us', i.e. objectivity within the limits of our knowledge, and not to be confused with the metaphysical realist's notion of objectivity as mind-independence.[5]

In more recent years, the debate on structural realism has revolved around the two camps of epistemological and ontic structural realists, where the latter have campaigned for 'a reconceptualisation of ontology, at the most basic metaphysical level, which effects a shift from objects to structures'.[6] Ontic structural realism draws significantly on Cassirer's neo-Kantianism in reversing the relationship between the concept of object and the concept of law, and taking the latter as expressing the structural relations 'in terms of which the "entities" are constituted.'[7]

The philosophical perspective on Pauli's exclusion principle that I have developed sits squarely within a Kantian realist position, even if throughout the book I have intentionally avoided engaging in discussions of 'truth' or 'realism', to focus instead on the rationale for the exclusion principle. I aimed to address an epistemological issue, namely under what conditions a contingent phenomenological rule could be accredited to the status of a scientific principle; I did not aim to offer an ontological thesis about what there is. Yet, as the redefinition of objectivity shows, from a Kantian viewpoint 'objective reality' is nothing but what we can have epistemic access to within the boundaries of a system of knowledge. A system that – as I have shown in this book – is built bottom-up, starting from phenomena to deduce phenomenological rules that can be embedded into larger mathematical structures, where their scope is redefined and they eventually acquire a crucial systematizing role for the deduction of other laws; until new puzzling phenomena are eventually discovered that may validate or invalidate them. No wonder the history of Pauli's principle is intertwined with the history of entities such as electrons, positrons, coloured quarks,

---

the internalist perspective, because it is characteristic of this view to hold that "what objects does the world consists of?" is a question that it only makes sense to ask *within* a theory or description.' Putnam (1981), pp. 49–50. Emphasis in the original.

[4] For a comparison between Putnam's internal realism and metaphysical realism in the light of Kant's philosophy of science, see Buchdahl (1986).

[5] 'What then is a true judgement? Kant does believe that we have *objective* knowledge: we know laws of mathematics, laws of geometry, laws of physics ... The use of the term "knowledge" and the use of the term "objective" amount to the assertion that *there is still a notion of truth*. But what is truth if it is not correspondence to the way things are in themselves? ... The only answer one can extract from Kant's writing is this: a piece of knowledge (i.e. a "true statement") is a statement that a rational being would accept on sufficient experience of the kind that it is actually possible for beings with our nature to have.' Putnam (1981), p. 64.

[6] French and Ladyman (2003a), p. 37.

[7] *Ibid.*, pp. 39–40. For criticism see Cao (2003a), (2003b); a response to Cao is in French and Ladyman (2003b).

and parons. As we saw in Chapter 5, a Kantian view leads us ultimately out of the regulative domain of systematization into the territory of empirical support. Phenomena retain the right of giving the final verdict to the system (and hence to the principles themselves), as is to be in an empirical and testable view of science. That is why the systematizing task of the exclusion principle is bound to remain open-ended. That is why, to echo Pauli's words at the beginning of this book, the history of the exclusion principle is already an old one, but its conclusion has not yet been written. The aim of this book was not to write a conclusion, but rather to show why it cannot be written.

# References

The abbreviation AHQP refers to the Archive for the History of Quantum Physics, whose content is listed in Kuhn, T. S. and Heilbron, J. L. *et al.* (eds.) (1967) *Sources for the History of Quantum Physics* (Philadelphia: American Philosophical Society). The numbers in the brackets refer to the microfilm number and to the section of the microfilm, respectively.

Pauli's primary sources cited are all reproduced in Pauli (1964).

Adler, S. L. (1969) 'Axial-vector vertex in spinor electrodynamics', *Physical Review* **177**, 2426–38.
  (2004) 'Anomalies to all orders', arXiv: hep-th/0405040.
  (2005) 'Remarks on the history of quantum chromodynamics', submitted to *Physics Today*. In arXiv: hep-ph/0412297.
d'Alembert, J. Le Rond (1751) 'Discours préliminaire des editeurs', in *Encyclopédie, ou Dictionnaire raisonné des sciences, des artes et des métiers, par une société de gens de lettres* (Paris: Briasson, Le Breton, Durand). English translation (1995) *Preliminary Discourse on the Encyclopaedia of Diderot*, translated by R. Schwab (Chicago: University of Chicago Press).
Allison, H. E. (1994) 'Causality and causal laws in Kant: a critique of Michael Friedman', in P. Parrini, (ed.) *Kant and Contemporary Epistemology* (Dordrecht: Kluwer Academic Publishers), 291–307.
Anderson, C. D. (1933) 'The positive electron', *Physical Review* **43**, 491–4.
Appelquist, T. and Politzer, H. D. (1975) 'Heavy quarks and $e^+e^-$ annihilation', *Physical Review Letters* **34**, 43–5.
Aubert, J. J., Ting, S. *et al.* (1974) 'Experimental observation of a heavy particle *J*', *Physical Review Letters* **33**, 1404–6.
Augustin, J. E., Richter, B. *et al.* (1974) 'Discovery of a narrow resonance in $e^+e^-$ annihilation', *Physical Review Letters* **33**, 1406–8.
Bacry, H. *et al.* (1964) 'Basic $SU_3$ triplets with integral charge and unit baryon number', *Physics Letters* **9**, 279–80.
Bardeen, W. A., Fritzsch, H., and Gell-Mann, M. (1973) 'Light-cone current algebra, $\pi^0$ decay, and $e^+e^-$ annihilation', in R. Gatto (ed.) *Scale and Conformal Symmetry in Hadron Physics* (New York: Wiley). Reissued in arXiv: hep-ph/0212183.

Barnes, V. E. et al. (1964) 'Observation of a hyperon with strangeness minus three', *Physical Review Letters* **12**, 204–6. Reprinted in M. Gell-Mann and Y. Ne'eman (eds.) (1964) *The Eightfold Way* (New York: W. A. Benjamin), 88–90.
Belinfante, F. J. (1939) 'The undor equation of the meson field', *Physica* **6**, 870–86.
Bell, J. S. and Jackiw, R. (1969) 'A PCAC puzzle: $\pi^0 \to \gamma\gamma$ in the $\sigma$-model', *Il Nuovo Cimento* **A60**, 47–61.
Bjorken, J. D. and Glashow, S. L. (1964) 'Elementary particles and SU(4)', *Physics Letters* **11**, 255–7.
Bjorken, J. D. and Paschos, E. A. (1969) 'Inelastic electron-proton and $\gamma$-proton scattering and the structure of nucleon', *Physical Review* **185**, 1975–82.
Blackett, P. M. S. and Occhialini, G. P. S. (1933) 'Some photographs of the tracks of penetrating radiation', *Proceedings of the Royal Society* **A139**, 699–726.
Bohr, N. (1913) 'On the constitution of atoms and molecules', *Philosophical Magazine* **26**, 1–25; 476–502; 857–75.
  (1914) 'Om Brintspektret', *Fysisk Tidsskrift* **12**, 97–114.
  (1923) 'Linienspektren und Atombau', *Annalen der Physik* **71**, 228–88.
  (1976) *Niels Bohr Collected Works*, Vol. 2, edited by L. Rosenfeld, J. Rud Nielsen et al. (Amsterdam: North-Holland Publishing Company).
  (1977) *Niels Bohr Collected Works*, Vol. 3–4, edited by L. Rosenfeld, J. Rud Nielsen et al. (Amsterdam: North-Holland Publishing Company).
Bohr, N. and Coster, D. (1923) 'Röntegenspektren und periodisches System der Elemente', *Zeitschrift für Physik* **12**, 342–74.
Bohr, N., Kramers, H. A., and Slater, J. C. (1924) 'The quantum theory of radiation', *Philosophical Magazine* **47**, 785–802.
Born, M. and Jordan, P. (1925) 'Zur Quantenmechanik' *Zeitschrift für Physik* **34**, 858–88.
Born, M., Heisenberg, W., and Jordan, P. (1926) 'Zur Quantenmechanik II' *Zeitschrift für Physik* **35**, 557–615.
Bose, S. N. (1924) 'Plancks Gesetz und Lichtquantenhypothese', *Zeitschrift für Physik* **26**, 178–81.
Brading, K. and Castellani, E. (eds.) (2003) *Symmetries in Physics* (Cambridge: Cambridge University Press).
Buchdahl, G. (1969a) *Metaphysics and the Philosophy of Science* (Cambridge, Mass.: MIT Press).
  (1969b) 'The Kantian "Dynamic of Reason", with special reference to the place of causality in Kant's system', in L. W. Beck (ed.) *Kant Studies Today* (La Salle, Ill.: Open Court), 341–74.
  (1974) 'The conception of lawlikeness in Kant's philosophy of science', in L. W. Beck (ed.) *Kant's Theory of Knowledge* (Dordrecht: Reidel), 128–50.
  (1986) 'Metaphysical and Internal Realism: the relations between ontology and methodology in Kant's philosophy of science', in R. Barcan Marcus et al. (eds.) *Logic, Methodology, and Philosophy of Science VII: Proceedings of the Seventh International Congress of Logic, Methodology and Philosophy of Science, Salzburg, 1983* (Amsterdam: North–Holland), 623–41.
Butts, R. E. (1991) 'Comments on Michael Friedman: "Regulative and Constitutive"', *The Southern Journal of Philosophy* **30**, Suppl., 103–8.

Cabibbo, N. (1963) 'Unitary symmetry and leptonic decay', *Physical Review Letters* **10**, 531–3.
Cao, T. (2003a) 'Structural realism and the interpretation of quantum field theory', *Synthese* **136**, 3–24.
  (2003b) 'Can we dissolve physical entities into mathematical structures?', *Synthese* **136**, 57–71.
Cassidy, D. C. (1979) 'Heisenberg's first core model of the atom: the formation of a professional style', *Historical Studies in Physical Sciences* **10**, 187–224.
Cassirer, E. (1910) *Substanzbegriff und Funktionsbegriff. Untersuchungen zu den Grundfragen der Erkenntniskritik* (Berlin: Bruno Cassirer). English translation (1953) *Substance and Function* and *Einstein's Theory of Relativity*, by W. C. Swabey (New York: Dover Publications).
  (1932) *Die Philosophie der Aufklärung* (Tübingen: J. C. B. Mohr). English translation, fifth edition (1962) *The Philosophy of the Enlightenment*, by F. Koelln and J. Pettegrove (Boston: Beacon Press).
  (1936) *Determinismus und Indeterminismus in der modernen Physik* (Göteborg: Högskolas Arsskrift 42). English translation (1956) *Determinism and Indeterminism in Modern Physics*, by O. T. Benfey (New Haven: Yale University Press).
Close, F. E. (1997) 'Glueballs and hybrids: new states of matter', *Contemporary Physics* **38**, 1–12.
Cowan, C. L., Reines, F. *et al.* (1956) 'Detection of the free neutrino: a confirmation', *Science* **124**, 103–4.
Darwin, C. G. (1927a) 'The electron as a vector wave', *Nature* **119**, 282–4.
  (1927b) 'The electron as a vector wave', *Proceedings of the Royal Society* **A116**, 227–53.
  (1928) 'The wave equation of the electron', *Proceedings of the Royal Society* **A118**, 654–80.
de Broglie, L. (1925) 'Recherche sur la théorie des quanta', *Annales de Physique* **3**, 22–128.
de Broglie, L. and Dauvillier, A. (1922) 'Sur les analogies de structure entre les séries optiques et les séries de Röntgen', *Comptes Rendus de l'Académie des Sciences, Paris* **175**, 755–6.
de Groot, J. G. H. *et al.* (1979) 'Inclusive interactions of high-energy neutrinos and antineutrinos in iron', *Zeitschrift für Physik* **C1**, 143–62.
Deilamian, K., Gillaspy, J. D., and Kelleher, D. E. (1995) 'Search for small violations of the symmetrization postulate in an excited state of helium', *Physical Review Letters* **74**, 4787–90.
Dirac, P. A. M. (1925) 'The fundamental equations of quantum mechanics' *Proceedings of the Royal Society* **A109**, 642–53.
  (1926a) 'Quantum mechanics and a preliminary investigation of the hydrogen atom' *Proceedings of the Royal Society* **A110**, 561–69.
  (1926b) 'On the theory of quantum mechanics', *Proceedings of the Royal Society* **A112**, 661–77.
  (1927) 'The quantum theory of the emission and absorption of radiation', *Proceedings of the Royal Society* **A114**, 243–65.
  (1928a) 'The quantum theory of the electron', *Proceedings of the Royal Society* **A117**, 610–24.

(1928b) 'The quantum theory of the electron. Part II', *Proceedings of the Royal Society* **A118**, 351–61.

(1930a) 'A theory of electrons and protons', *Proceedings of the Royal Society* **A126**, 360–5.

(1930b) 'On the annihilation of electrons and protons', *Proceedings of the Cambridge Philosophical Society* **26**, 361–75.

(1931) 'Quantised singularities in the electromagnetic field', *Proceedings of the Royal Society* **A133**, 60–72.

(1934a) 'Theory of the positron', in *Structure et Propriétés des Noyaux Atomiques. Rapports et discussions du septième conseil de physique tenue à Bruxelles du 22 au 29 Octobre 1933 sous les auspices de l'Institut International de Physique Solvay* (Paris: Gauthier-Villars), 203–30.

(1934b) 'Discussion of the infinite distribution of electrons in the theory of the positron', *Proceedings of the Cambridge Philosophical Society* **30**, 150–63.

Dorling, J. (1973) 'Demonstrative induction: its significant role in the history of physics', *Philosophy of Science* **49**, 360–72.

(1974) 'Henry Cavendish's deduction of the electrostatic inverse square law from the result of a single experiment', *Studies in the History and Philosophy of Science* **4**, 327–48.

(1991) 'Reasoning from phenomena: lessons from Newton', *PSA 1990*, vol. 2, 197–208.

Drake, G. W. F. (1989) 'Predicted energy shift for paronic helium', *Physical Review* **A39**, 897–9.

Duck, I. and Sudarshan, E. C. G. (1997) *Pauli and the Spin-Statistics Theorem* (Singapore: World Scientific).

Duhem, P. (1906) *La Theorie Physique: Son Object, Sa Structure* (Paris: Marcel Riviere & Cie.). English translation (1991) *The Aim and Structure of Physical Theory* (Princeton: Princeton University Press).

Dyson, F. J. (1967) 'Ground-state energy of a finite system of charged particles', *Journal of Mathematical Physics* **8**, 1538–45.

(1996) *Selected Papers of Freeman Dyson with commentary* (Cambridge, Mass.: International Press).

Dyson, F. J. and Lenard, A. (1967) 'Stability of matter I', *Journal of Mathematical Physics* **8**, 423–34.

(1968) 'Stability of matter II', *Journal of Mathematical Physics* **9**, 698–711.

Eco, U. (1984) *Semiotics and the Philosophy of Language* (Bloomington: Indiana University Press).

Ehrenfest, P. (1913) 'Een mechanische theorema van Boltzmann en zijne betrekking tot de quanta theorie', *Verslag van de Gewoge Vergaderingen der Wis-en Natuurkundinge Afdeeling*, Amsterdam, 586–93. English translation (1914) 'A mechanical theorem of Boltzmann and its relation to theory of energy quanta', *Proceedings of the Amsterdam Academy* **16**, 591–7.

(1927) 'Besteht ein allgemeiner Zusammenhang zwischen der wechselseitigen Undurchdringlichkeit materieller Teilchen und dem "Pauli-Verbot"?' *Naturwissenschaften* **15**, 161–2.

Einstein, A. (1924) 'Quantentheorie des einatomigen idealen Gases', *Sitzungsberichte Preussische Akademie der Wissenschaften, Physikalisch–Mathematische Klasse*, 261–67.
  (1925a) 'Quantentheorie des einatomigen idealen Gases. 2. Abhandlung', *Sitzungsberichte Preussische Akademie der Wissenschaften, Physikalisch–Mathematische Klasse*, 3–14.
  (1925b) 'Quantentheorie des idealen Gases', *Sitzungsberichte Preussische Akademie der Wissenschaften, Phyisikalisch–Mathematische Klasse*, 18–25.
Enz, C. P. (2002) *No Time to be Brief. A Scientific Biography of Wolfgang Pauli* (Oxford: Oxford University Press).
Fermi, E. (1926) 'Sulla quantizzazione del gas perfetto monoatomico', *Rendiconti della Reale Accademia dei Lincei* **3**, 145–9.
  (1934) 'Versuch einer Theorie der β–Strahlen', *Zeitschrift für Physik* **88**, 161–71.
  (1962) *Collected Papers*, Vol. I (Chicago: University of Chicago Press).
Feynman, R. P. (1972), *Photon-Hadron Interactions* (New York: Benjamin).
Feynman, R. P. and Gell-Mann, M. (1958) 'Theory of the Fermi interaction', *Physical Review* **109**, 193–98.
Fierz, M. (1939) 'Über die relativistische Theorie kräftefreier Teilchen mit beliebigem Spin', *Helvetica Physica Acta* **12**, 3–37.
Fierz, M. and Pauli, W. (1939) 'Über relativistische Feldgleichungen von Teilchen mit beliebigem Spin im elektromagnetischen Feld', *Helvetica Physica Acta* **12**, 297–300. Reprinted with the English translation (1939) in *Proceedings of the Royal Society* **A173**, 211–32.
Forman, P. (1968) 'The doublet riddle and atomic physics *circa* 1924', *Isis* **59**, 156–74.
  (1970) 'Alfred Landé and the anomalous Zeeman effect, 1919–1921', *Historical Studies in the Physical Sciences* **2**, 153–261.
French, S. (1984) 'Identity and individuality in classical and quantum physics', Ph.D. thesis, University of London.
  (1995) 'The esperable uberty of quantum chromodynamics', *Studies in History and Philosophy of Modern Physics* **26**, 87–105.
French, S. and Ladyman, J. (2003a) 'Remodelling structural realism: quantum physics and the metaphysics of structure', *Synthese* **136**, 31–56.
  (2003b) 'The dissolution of objects: between Platonism and Phenomenalism', *Synthese* **136**, 73–7.
French, S. and Redhead, M. (1988) 'Quantum physics and the identity of indiscernibles', *British Journal for the Philosophy of Science* **39**, 233–46.
French, S. and Rickles, D. (2003) 'Understanding permutation symmetry', in K. Brading and E. Castellani (eds.) *Symmetries in Physics* (Cambridge: Cambridge University Press), 212–38.
Frenkel, J. (1926) 'Die Elektrodynamik des rotierenden Elektrons', *Zeitschrift für Physik* **37**, 243–62.
Friedman, M. (1989) 'Kant on space, the understanding, and the law of gravitation: *Prolegomena* 38', *Monist* **79**, 236–84.
  (1991) 'Regulative and constitutive', *The Southern Journal of Philosophy* **30**, Suppl., 73–102.

(1992a) *Kant and the Exact Sciences* (Cambridge, Mass.: Harvard University Press).
  (1992b) 'Causal laws and the foundations of natural science', in P. Guyer (ed.) *The Cambridge Companion to Kant* (Cambridge: Cambridge University Press), 161–99.
  (1994) 'Kant and the twentieth century', in P. Parrini (ed.) *Kant and Contemporary Epistemology* (Dordrecht: Kluwer Academic Publishers), 27–46.
  (1999) *Reconsidering Logical Positivism* (Cambridge: Cambridge University Press).
  (2000a) 'Transcendental philosophy and a priori knowledge: a neo-Kantian perspective', in P. Boghossian and C. Peacocke (eds.) *New Essays on the A Priori* (Oxford: Clarendon Press).
  (2000b) *A Parting of the Ways. Carnap, Cassirer, Heidegger* (La Salle, Ill.: Open Court).
  (2001) *The Dynamics of Reason. Stanford Kant Lectures* (Stanford: CSLI Publications).
Fritzsch, H. and Gell-Mann, M. (1972) 'Light cone current algebra', hep-ph/0301127.
Galison, P. (1997) *Image and Logic* (Chicago: University of Chicago Press).
Gell-Mann, M. (1961) 'The Eightfold Way: A Theory of Strong Interaction Symmetry', California Institute of Technology Synchrotron Laboratory Report CTSL-20. Reprinted in M. Gell-Mann and Y. Ne'eman (eds.) (1964) *The Eightfold Way* (New York: W. A. Benjamin), 11–57.
  (1962a) 'Strange particle physics. Strong interactions', *Proceedings of the International Conference on High Enery Physics* (CERN, 1962). Reprinted in M. Gell-Mann and Y. Ne'eman (eds.) (1964) *The Eightfold Way* (New York: W. A. Benjamin), 87.
  (1962b) 'Symmetries of baryons and mesons', *Physical Review* **125**, 1067–84.
  (1964) 'A schematic model of baryons and mesons', *Physics Letters* **8**, 214–5. Reprinted in M. Gell-Mann and Y. Ne'eman (eds.) (1964) *The Eightfold Way* (New York: W. A. Benjamin), 168–9.
Gell-Mann, M. and Lévy, M. (1960) 'The axial vector current in beta decay', *Il Nuovo Cimento* **16**, 705–25.
Gentile, G. (1940) 'Osservazioni sopra le statistiche intermedie', *Il Nuovo Cimento* **17**, 493–7.
Glashow, S. L. (1961) 'Partial symmetries of weak interactions', *Nuclear Physics* **22**, 579–88.
Glashow, S. L., Iliopoulos, J., and Maiani, L. (1970) 'Weak interactions with lepton-hadron symmetry', *Physical Review* **D2**, 1285–92.
Goldhaber, M. and Scharff-Goldhaber, G. (1948) 'Identification of beta-rays with atomic electrons', *Physical Review* **73**, 1472–3.
Gordon, W. (1926) 'Der Compton Effekt nach der Schrödingerschen Theorie', *Zeitschrift für Physik* **40**, 117–33.
  (1928) 'Die Energieniveaus des Wasserstoffatoms nach der Diracschen Quantentheorie des Elektrons', *Zeitschrift für Physik* **48**, 11–4.
Green, H. S. (1953) 'A generalised method of field quantization', *Physical Review* **90**, 270–3.

Greenberg, O. W. (1964) 'Spin and unitary-spin independence in a paraquark model of baryons and mesons' *Physical Review Letters* **13**, 598–602.
  (1991) 'Particles with small violations of Fermi or Bose statistics', *Physical Review* **D43**, 4111–20.
  (1993) 'Color: from baryon spectroscopy to QCD', in Moshe Gai (ed.) *International Conference on the Structure of Baryons and Related Mesons, June 1–4, 1992, Yale University* (Singapore; River Edge, NJ: World Scientific).
  (1999) 'Quon statistics for composite systems and a limit on the violation of the Pauli principle for nucleons and quarks', *Physical Review Letters* **83**, 4460–63.
Greenberg, O. W. and Macrae, K. I. (1983) 'Locally gauge-invariant formulation of parastatistics', *Nuclear Physics* **B219**, 358–66.
Greenberg, O. W. and Mohapatra, R. N. (1989) 'Phenomenology of small violations of Fermi and Bose statistics', *Physical Review* **D39**, 2032–8.
Greenberg, O. W. and Nelson, C. A. (1977) 'Color models of hadrons', *Physics Reports* **32**, 69–121.
Gross, D. J. and Wilczek, F. (1973) 'Asymptotically free gauge theories: I', *Physical Review* **D8**, 3633–52.
Guyer, P. (1990) 'Reason and reflective judgement: Kant on the significance of systematicity', *Noûs* **24**, 17–43.
  (2003) 'Kant's principles of reflecting judgment', in P. Guyer (ed.) *Kant's Critique of the Power of Judgment. Critical Essays* (Lanham: Rowman & Littlefield Publishers), 1–61.
ter Haar, D. (1952) 'Gentile's intermediate statistics', *Physica* **18**, 199–200.
Hacking, I. (1983) *Representing and Intervening* (Cambridge: Cambridge University Press).
  (1993) 'Working in a new world: the taxonomic solution', in P. Horwich (ed.) *World Changes. Thomas Kuhn and the Nature of Science* (Cambridge, Mass.: MIT Press), 275–310.
Han, M. Y. and Nambu, Y. (1965) 'Three-triplet model with double SU(3) symmetry', *Physical Review* **139**, B1006–10.
Harper, W. (1990) 'Newton's classic deductions from phenomena', *PSA 1990: Proceedings of the 1990 Biennial Meeting of the Philosophy of Science Association*, vol. 2 (East Lansing, Mich.: Philosophy of Science Association).
Harper, W. and Smith, G. E. (1995) 'Newton's new way of inquiry', in J. Leplin (ed.) *The Creation of Ideas in Physics* (Dordrecht: Kluwer), 113–66.
Heilbron, J. L. (1966) 'The work of H. G. J. Moseley', *Isis* **57**, 336–64.
  (1967) 'The Kossel-Sommerfeld theory and the ring atom', *Isis* **58**, 451–85.
  (1982) 'The origins of the exclusion principle', *Historical Studies in the Physical Sciences* **13**, 261–310.
Heisenberg, W. (1922) 'Zur Quantentheorie der Linienstruktur und der anomalen Zeemaneffekte', *Zeitschrift für Physik* **8**, 273–97.
  (1924) 'Über eine Abänderung der formalen Regeln der Quantentheorie beim Problem der anomalen Zeemaneffekte', *Zeitschrift für Physik* **26**, 291–307.
  (1925a) 'Zur Quantentheorie der Multiplettstruktur und der anomalen Zeemaneffekte', *Zeitschrift für Physik* **32**, 841–60.
  (1925b) 'Über quantentheoretische Umdeutung kinematischer und mechanischer Beziehungen', *Zeitschrift für Physik* **33**, 879–93.

(1926) 'Über die Spektra von Atomsystemen mit zwei Elektronen', *Zeitschrift für Physik* **39**, 499–518.

(1934) 'Bemerkungen zur Diracschen Theorie des Positrons', *Zeitschrift für Physik* **90**, 209–31.

Heisenberg, W. and Jordan, P. (1926) 'Anwendung der Quantenmechanik auf das Problem der anomalen Zeemaneffekte', *Zeitschrift für Physik* **37**, 263–77.

Heisenberg, W. and Pauli, W. (1929) 'Zur Quantenelektodynamik der Wellenfelder', *Zeitschrift für Physik* **56**, 1–61.

(1930) 'Zur Quantentheorie der Wellenfelder II', *Zeitschrift für Physik* **59**, 168–90.

Hesse, M. (1983) 'Comment on Kuhn's "Commensurability, Comparability, Communicability"', in P. D. Asquith and T. Nickles (eds.) *PSA 1982: Proceedings of the 1982 Biennial Meeting of the Philosophy of Science Association*, vol. 2 (East Lansing, Mich.: Philosophy of Science Association), 707–11.

Higgs, P. W. (1964a) 'Broken symmetries, massless particles, and gauge fields', *Physics Letters* **12**, 132–3.

(1964b) 'Broken symmetries and the masses of gauge bosons', *Physical Review Letters* **13**, 508–9.

Hilborn, R. C. and Tino, G. M. (eds) (2000) *Spin-Statistics Connection and Commutation Relations: Experimental Tests and Theoretical Implications* (Melville, NY: American Institute of Physics).

Hofmann, W. (1981) *Jets of Hadrons* (Berlin: Springer Verlag).

Jackiw, R. (1999) 'The unreasonable effectiveness of quantum field theory', in T. Y. Cao (ed.) *Conceptual Foundations of Quantum Field Theory* (Cambridge: Cambridge University Press), 148–59.

Jammer, M. (1966) *The Conceptual Development of Quantum Mechanics* (New York: McGraw Hill); second edition (1989) (American Institute of Physics: Tomash Publishers).

Jordan, P. (1925) 'Bemerkungen zur Theorie der Atomstruktur', *Zeitschrift für Physik* **33**, 563–70.

(1927) 'Zur Quantenmechanik der Gasentartung', *Zeitschrift für Physik* **44**, 473–80.

Jordan, P. and Pauli, W. (1928) 'Zur Quantenelektrodynamik ladungsfreier Felder', *Zeitschrift für Physik* **47**, 151–73.

Jordan, P. and Wigner E. (1928) 'Über das Paulische Äquivalenzverbot', *Zeitschrift für Physik* **47**, 631–51.

Kant, I. (1781) *Critik der reinen Vernunft* (Riga: Johann Hartknoch). English translation (1997) P. Guyer and A. W. Wood (eds.) *Critique of Pure Reason* (Cambridge: Cambridge University Press).

(1790) *Kritik der Urteilskraft* (Berlin: Lagarde). English translation (2000) P. Guyer and E. Matthews (eds.) *Critique of the Power of Judgment* (Cambridge: Cambridge University Press).

Kitcher, P. (1983) 'Implications of incommensurability', in P. D. Asquith and T. Nickles (eds.) *PSA 1982: Proceedings of the 1982 Biennial Meeting of the Philosophy of Science Association*, vol. 2 (East Lansing, Mich.: Philosophy of Science Association), 692–3.

(1986) 'Projecting the order of nature', in R. E. Butts (ed.) *Kant's Philosophy of Physical Science* (Dordrecht: Reidel), 201–35.

Klein, O. (1927) 'Elektrodynamik und Wellenmechanik von Standpunkt des Korrespondenzprinzips', *Zeitschrift für Physik* **41**, 407–42.

Klein, O. and Nishina Y. (1929) 'Über die Streuung von Strahlung durch freie Elektronen nach der neuen relativistischen Quantendynamik von Dirac', *Zeitschrift für Physik* **52**, 853–68.

Kragh, H. (1979) 'Niels Bohr's second atomic theory', *Historical Studies in the Physical Sciences* **10**, 123–86.

Kripke, S. (1972) 'Naming and necessity', in G. Harman and D. Davidson (eds.) *The Semantics of Natural Language* (Dordrecht: Reidel).

Kronig, R. (1960) 'The turning point', in M. Fierz and V. F. Weisskopf (eds.) *Theoretical Physics in the Twentieth Century: A Memorial Volume to Wolfgang Pauli* (New York: Interscience Publishers).

Kuhn, T. S. (1957) *The Copernican Revolution: Planetary Astronomy in the Development of Western Thought* (Cambridge, Mass.: Harvard University Press).

(1962) *The Structure of Scientific Revolutions*, International Encyclopaedia of Unified Science: Foundations of the Unity of Science, vol. 2, no. 2 (Chicago: University of Chicago Press).

(1970) 'Reflections on my critics', in I. Lakatos and A. Musgrave (eds.) *Criticism and the Growth of Knowledge: Proceedings of the International Colloquium in the Philosophy of Science*, London 1965, vol. IV (Cambridge: Cambridge University Press), 231–78. Reprinted in Kuhn (2000), 123–75.

(1977) *The Essential Tension: Selected Studies in Scientific Tradition and Change* (Chicago: University of Chicago Press).

(1978) *Black-Body Theory and the Quantum Discontinuity 1894–1912* (Oxford: Oxford University Press).

(1983) 'Commensurability, comparability, communicability', in P. D. Asquith and T. Nickles (eds.) *PSA 1982: Proceedings of the 1982 Biennial Meeting of the Philosophy of Science Association*, vol. 2 (East Lansing, Mich.: Philosophy of Science Association), 669–88. Reprinted in Kuhn (2000), 33–58.

(1989) 'Possible worlds in history of science', in Sture Allén (ed.) *Possible Worlds in Humanities, Arts and Sciences: Proceedings of Nobel Symposium 65* (Berlin: Walter de Gruyter). Reprinted in Kuhn (2000), 58–90.

(1990) 'Dubbing and redubbing: the vulnerability of rigid designation', in C. Wade Savage (ed.) *Scientific Theories*, Minnesota Studies in the Philosophy of Science **14** (Minneapolis: University of Minnesota Press), 298–318.

(1991) 'The road since structure', in A. Fine, M. Forbes, and L. Wessels (eds.) *PSA 1990: Proceedings of the 1990 Biennial Meeting of the Philosophy of Science Association*, vol. 2 (East Lansing, Mich.: Philosophy of Science Association), 3–13. Reprinted in Kuhn (2000), 90–105.

(1993) 'Afterwards', in P. Horwich (ed.) *World Changes. Thomas Kuhn and the Nature of Science* (Cambridge, Mass.: MIT Press), 311–39. Reprinted in Kuhn (2000), 224–53.

(2000) *The Road Since Structure. Philosophical Essays, 1970–1993, with an Autobiographical Interview* (Chicago: University of Chicago Press).

Lakatos, I. (1978) *Philosophical Papers, Vol. I: The Methodology of Scientific Research Programmes* (Cambridge: Cambridge University Press).
Landé, A. (1921a) 'Über den anomalen Zeemaneffekt (Teil I)', *Zeitschrift für Physik* **5**, 231–41.
  (1921b) 'Über den anomalen Zeemaneffekt (Teil II)', *Zeitschrift für Physik* **7**, 398–405.
  (1923a) 'Termstruktur und Zeemaneffekt der Multipletts', *Zeitschrift für Physik* **15**, 189–205; 'Termstruktur und Zeemaneffekt der Multipletts. Zweite Mitteilung', *Zeitschrift für Physik* **19**, 112–23.
  (1923b) 'Zur Theorie der Röntgenspektren', *Zeitschrift für Physik* **16**, 391–96.
  (1923c) 'Zur Struktur des Neonspektrums', *Zeitschrift für Physik* **17**, 292–4.
  (1923d) 'Feinstruktur und Zeemaneffekt der Multipletts', *Zeitschrift für Physik* **19**, 112–23.
Landé, A. and Heisenberg, W. (1924) 'Termstruktur der Multipletts höherer Stufe', *Zeitschrift für Physik* **25**, 279–86.
Larmor, J. (1897) 'On the theory of the magnetic influence on spectra; and on the radiation from moving ions', *Philosophical Magazine* **44**, 503–12.
Laudan, L. (1990) 'Demistifying underdetermination in scientific theories', in C. Wade Savage (ed.) *Scientific Theories,* Minnesota Studies in the Philosophy of Science **14** (Minneapolis: University of Minnesota Press), 267–97.
Lieb, E. H. (1991) *The Stability of Matter: From Atoms to Stars. Selecta of Elliott H. Lieb* (Berlin: Springer-Verlag). Second edition(1997).
Lieb, E. H. and Thirring, W. E. (1991) 'Bound for the kinetic energy of fermions which proves the stability of matter', *Physical Review Letters* **35**, 687–90.
Lipton, P. (2001) 'Kant on wheels', *London Review of Books*, 19 July, 30–1. Reprinted in *Social Epistemology* **17**, 2003, 215–19.
Lorentz, H. A. (1897) 'Über den Einfluss magnetischer Kräfte auf die Emission des Lichtes', *Wiedemannsche Annalen der Physik* **63**, 278–84.
Majorana, E. (1932) 'Teoria relativistica di particelle con momento intrinseco arbitrario', *Il Nuovo Cimento* **9**, 335–44.
Massimi, M. (2001) 'Exclusion principle and the identity of indiscernibles: a response to Margenau's argument', *British Journal for the Philosophy of Science* **52**, 303–31.
  (2004a) 'Non-defensible middle ground for experimental realism: why we are justified to believe in colored quarks', *Philosophy of Science* **71**, 36–60.
  (2004b) 'What demonstrative induction can do against the threat of underdetermination: Bohr, Heisenberg, and Pauli on spectroscopic anomalies (1921–24)', *Synthese* **140**, 243–77.
Massimi, M. and Redhead, M. (2003) 'Weinberg's proof of the spin–statistics theorem', *Studies in History and Philosophy of Modern Physics* **34**, 621–50.
Mehra, J. and Rechenberg, H. (1982a) *The Historical Development of Quantum Theory. Vol. 2: The Discovery of Quantum Mechanics 1925* (New York: Springer Verlag).
  (1982b) *The Historical Development of Quantum Theory. Vol. 3: The Formulation of Matrix Mechanics and its Modifications, 1925–1926* (New York: Springer Verlag).

(1982c) *The Historical Development of Quantum Theory. Vol. 4, Part 1: The Fundamental Equations of Quantum Mechanics, 1925–1926. Part 2: The Reception of the New Quantum Mechanics, 1925–1926* (New York: Springer Verlag).
Messiah, A. M. L. and Greenberg, O. W. (1964a) 'Symmetrization postulate and its experimental foundation', *Physical Review* **136**, B248–67.
  (1964b) 'Selection rules for parafields and the absence of para particles in nature' *Physical Review* **138**, B1155–67.
Miyamoto, Y. (1965) 'Three kinds of triplet model', in *Extra Number Supplement of Progress of Theoretical Physics: Thirtieth Anniversary of the Yukawa Meson Theory*, p. 187.
Ne'eman, Y. (1961) 'Derivation of strong interactions from a gauge invariance', *Nuclear Physics* **26**, 222–9. Reprinted in M. Gell-Mann and Y. Ne'eman (eds.) (1964) *The Eightfold Way* (New York: W. A. Benjamin), 58–65.
Ne'eman, Y. and Kirsh, Y. (1986) *The Particle Hunters* (Cambridge: Cambridge University Press).
Newton, I. (1687) *Philosophiae Naturalis Principia Mathematica*. English translation (1803) *The Mathematical Principles of Natural Philosophy*, by W. Davis (London: H. D. Symonds).
Nickles, T. (ed.) (2003) *Thomas Kuhn* (Cambridge: Cambridge University Press).
Norton, J. D. (1993) 'The determination of theory by evidence: the case for quantum discontinuity', *Synthese* **97**, 1–31.
  (1994) 'Science and certainty', *Synthese* **99**, 3–22.
  (1995) 'Eliminative induction as a method of discovery: how Einstein discovered general relativity', in J. Leplin (ed.) *The Creation of Ideas in Physics* (Dordrecht: Kluwer).
Ohnuki, Y. and Kamefuchi, S. (1982) *Quantum Field Theory and Parastatistics* (Berlin: Springer).
Oppenheimer, J. R. (1930) 'Two notes on the probability of radiative transitions', *Physical Review* **35**, 939–47.
Oppenheimer, J. R. and Plesset, M. S. (1933) 'On the production of the positive electron', *Physical Review* **44**, 53–5.
Paschen, F. and Back, E. (1921) 'Liniengruppen magnetish vervollständigt', *Physica* **1**, 261–73.
Pauli, W. (1923) 'Über die Gesetzmäßigkeiten des anomalen Zeemaneffektes', *Zeitschrift für Physik* **16**, 155–64.
  (1924) 'Zur Frage der Zuordnung der Komplexstrukturterme in starken und in schwachen äußeren Feldern', *Zeitschrift für Physik* **20**, 371–87.
  (1925a) 'Über den Einfluß der Geschwindigkeitsabhängigkeit der Elektronenmasse auf den Zeemaneffekt', *Zeitschrift für Physik* **31**, 373–85.
  (1925b) 'Über den Zusammenhang des Abschlusses der Elektronengruppen im Atom mit der Komplexstruktur der Spektren', *Zeitschrift für Physik* **31**, 765–83.
  (1926a) 'Quantentheorie', in H. Geiger and K. Scheel (eds.) *Handbuch der Physik* (Berlin: Springer), vol. 23, 1–278.
  (1926b) 'Über das Wasserstoffspektrum vom Standpunkt der neuen Quantenmechanik', *Zeitschrift für Physik* **36**, 336–63.

(1927a) 'Über Gasentartung und Paramagnetismus', *Zeitschrift für Physik* **41**, 81–102.

(1927b) 'Zur Quantenmechanik des magnetischen Elektrons', *Zeitschrift für Physik* **43**, 601–23.

(1936) 'Théorie quantique relativiste des particules obéissant à la statistique de Einstein-Bose', *Annales de l'Institut Henri Poincaré* **6**, 137–52.

(1940) 'The connection between spin and statistics', *Physical Review* **58**, 716–22.

(1946) 'Remarks on the history of the exclusion principle', *Science* **103**, 213–15.

(1948) 'Exclusion principle and quantum mechanics' in *Les Prix Nobel en 1946* (Stockholm: Norstedt & Söner), 131–47.

(1955) 'Rydberg and the periodic system of elements', *Proceedings of the Rydberg Centennial Conference on Atomic Spectroscopy, Lund 1954, Universitets Arsskrift* **50**, 22–6.

(1964) *Collected Scientific Papers by Wolfgang Pauli*, Vols. I–II, edited by R. Kronig and V. F. Weisskopf (New York, London: Wiley Interscience).

(1979) *Wissenschaftlicher Briefwechsel mit Bohr, Einstein, Heisenberg u.a., Band 1: 1919–1929*, edited by A. Hermann, K. von Meyenn, and V. F. Weisskopf (Berlin, Heidelberg: Springer Verlag).

(1985) *Wissenschaftlicher Briefwechsel mit Bohr, Einstein, Heisenberg u.a., Band 2: 1930–1939*, edited by K. von Meyenn (Berlin: Springer Verlag).

(1993) *Wissenschaftlicher Briefwechsel mit Bohr, Einstein, Heisenberg u.a., Band 3: 1940–1949*, edited by K. von Meyenn (Berlin: Springer Verlag).

Pauli, W. and Belinfante, F. J. (1940) 'On the statistical behaviour of known and unknown elementary particles', *Physica* **7**, 177–92.

Pauli, W. and Weisskopf, V. (1934) 'Über die Quantisierung der skalaren relativistischen Wellengleichung', *Helvetica Physica Acta* **7**, 709–31.

Petruccioli, S. (1988) *Atomi metafore paradossi* (Roma: Theoria). English translation (1993) *Atoms, Metaphors and Paradoxes: Niels Bohr and the Construction of a New Physics* (Cambridge: Cambridge University Press).

Pickering, A. (1984) *Constructing Quarks* (Edinburgh: Edinburgh University Press).

Poincaré, H. (1902) *La science et l'hypothèse* (Paris: Flammarion). English translation (1982) 'Science and hypothesis', in J. Royce (ed.) *The Foundations of Science* (Washington: University Press of America).

(1905) *La valeur de la science* (Paris: Flammarion). English translation (1982) 'The value of science,' in J. Royce (ed.) *The Foundations of Science* (Washington: University Press of America).

Politzer, H. D. (1973) 'Reliable perturbative results for strong interactions?', *Physical Review Letters* **30**, 1346–9.

Popper, K. (1934) *Logik der Forschung* (Vienna: Springer). English translation (1968) *The Logic of Scientific Discovery* (London: Hutchinson & Co.).

(1972) *Objective Knowledge* (Oxford: Clarendon Press).

Putnam, H. (1975) 'The meaning of "meaning"', in K. Gunderson (ed.) *Language, Mind and Knowledge*, Minnesota Studies in the Philosophy of Science **7** (Minneapolis: University of Minnesota Press). Reprinted in H. Putnam (1975) *Mind, Language and Reality. Philosophical Papers*, Vol. 2 (Cambridge: Cambridge University Press), 215–71.

(1981) *Reason, Truth and History* (Cambridge: Cambridge University Press).

Quine, W. V. O. (1951) 'Two dogmas of empiricism', *Philosophical Review* **60**, 20–43.
  (1953) *From a Logical Point of View* (Cambridge, Mass.: Harvard University Press).
Ramberg, E. and Snow, G. (1990) 'Experimental limit on a small violation of the Pauli principle', *Physics Letters* **B238**, 438–41.
Redhead, M. (1975) 'Symmetry in intertheory relations', *Synthese* **32**, 77–112.
  (1982) 'Quantum field theory for philosophers', *PSA: Proceedings of the 1982 Biennial Meeting of the Philosophy of Science Association*, vol. 2 (East Lansing, Mich.: Philosophy of Science Association), pp. 57–99.
  (2003) 'The interpretation of gauge symmetry', in K. Brading and E. Castellani (eds.) *Symmetries in Physics* (Cambridge: Cambridge University Press), 124–39.
Redhead, M. and Teller, P. (1992) 'Particle labels and the theory of indistinguishable particles in quantum mechanics', *British Journal for the Philosophy of Science* **43**, 201–18.
Reichenbach, H. (1920) *Relativitätstheorie und Erkenntnis Apriori* (Berlin: Springer). English translation (1965) *The Theory of Relativity and A Priori Knowledge* (Los Angeles: University of California Press).
Richter, B. (1977) 'From the psi to charm: the experiments of 1975 and 1976', *Review of Modern Physics* **49**, 251–66.
Rubinowicz, A. (1918) 'Borsche Frequenzbedingung und Erhaltung des Impulsmomentes. Teil I', *Physikal Zeitschrift* **19**, 441–5.
Runge, C. (1907) 'Über die Zerlegung von Spektallinien im magnetischen Felde', *Physikalische Zeitschrift* **8**, 232–37.
Salam, A. and Ward, J. C. (1964) 'Electromagnetic and weak interactions', *Physics Letters* **13**, 168–71.
Schrödinger, E. (1926a) 'Quantisierung als Eigenwertproblem. Erste Mitteilung', *Annalen der Physik* **79**, 361–76.
  (1926b) 'Quantisierung als Eigenwertproblem. Zweite Mitteilung', *Annalen der Physik* **79**, 489–527.
  (1926c) 'Über das Verhältnis der Heisenberg–Born–Jordanschen Quantenmechanik zu der meinen', *Annalen der Physik* **79**, 734–56.
Serwer, D. (1977) '*Unmechanischer Zwang:* Pauli, Heisenberg, and the rejection of the mechanical atom 1923–1925', *Historical Studies in the Physical Sciences* **8**, 189–256.
Silvestrini, V. (1972) 'Electron-positron interactions', in J. D. Jackson and A. Roberts (eds.) *Proceedings of the XVI International Conference on High Energy Physics*, National Accelerator Laboratory, Batavia, Illinois, 6–13 September 1972 (Batavia: National Accelerator Laboratory).
Slater, J. C. (1929) 'The theory of complex spectra', *Physical Review* **34**, 1293–322.
Sommerfeld, A. (1916a) 'Zur Quantentheorie der Spektrallinien', *Annalen der Physik* **51**, 1–94; 125–67.
  (1916b) 'Zur Theorie des Zeeman-Effekts der Wasserstofflinien, mit einem Anhang über den Stark-Effekt' *Physikalische Zeitschrift* **17**, 491–507.
  (1919) *Atombau und Spektrallinien* (Braunschweig: Vieweg).
  (1920) 'Allgemeine spektroskopische Gesetze, insbesondere ein magnetooptischer Zerlegungssatz', *Annalen der Physik* **63**, 221–63.

(1922) 'Quantentheoretische Umdeutung der Voigtschen Theorie des anomalen Zeemaneffektes vom D-Linientypus', *Zeitschrift für Physik* **8**, 257–72.

Stolt, R. H. and Taylor, J. R. (1970) 'Correspondence between the first- and the second-quantized theories of paraparticles', *Nuclear Physics* **19B**, 1–19.

Stoner, E. C. (1924) 'The distribution of electrons among atomic levels', *Philosophical Magazine* **48**, 719–36.

Tavkhelidze, A. (1965) 'Higher symmetries and composite models of elementary particles', in *High Energy Physics and Elementary Particles* (Vienna: International Atomic Energy Agency), 753–62.

Thomas, L. H. (1926) 'The motion of the spinning electron', *Nature* **117**, 514.

(1927) 'The kinematics of an electron with an axis', *Philosophical Magazine* (7)**3**, 1–22.

't Hooft, G. (1971) 'Renormalizable Lagrangians for massive Yang-Mills fields', *Nuclear Physics* **B35**, 167–88.

Tomonaga, S. (1997) *The History of Spin* (Chicago: University of Chicago Press).

Uhlenbeck, G. E. and Goudsmit, S. (1925) 'Ersetzung der Hypothese vom unmechanischer Zwang durch eine Forderung bezüglich des inneren Verhaltens jedes einzelnen Elektrons', *Naturwissenschaften* **13**, 953–4.

(1926) 'Spinning electrons and the structure of spectra', *Nature* **117**, 264–5.

Van der Waerden, B. L. (1960) 'Exclusion principle and spin', in M. Fierz and V. F. Weisskopf (eds.) *Theoretical Physics in the Twentieth Century: A Memorial Volume to Wolfgang Pauli* (New York: Interscience Publishers), 199–244.

Van Fraassen, Bas (1980) *The Scientific Image* (Oxford: Oxford University Press).

(2002) *The Empirical Stance* (New Haven and London: Yale University Press).

Van Vleck J. H. (1922) 'The normal helium atom and its relation to the quantum theory' *Philosophical Magazine* **44**, 842–69.

Von Neumann, J. and Wigner, E. (1928a) 'Zur Erklärung einiger Eigenschaften der Spektren aus der Quantenmechanik des Drehelektrons', *Zeitschrift für Physik* **47**, 203–20.

(1928b) 'Zur Erklärung einiger Eigenschaften der Spektren aus der Quantenmechanik des Drehelektrons. Zweiter Teil', *Zeitschrift für Physik* **49**, 73–94.

(1928c) 'Zur Erklärung einiger Eigenschaften der Spektren aus der Quantenmechanik des Drehelektrons. Dritter Teil', *Zeitschrift für Physik* **51**, 844–58.

Weinberg, S. (1964) 'Feynman Rules for any spin', *Physical Review* **133**, B1318–32.

Weisskopf, V. (1934) 'Über die Selbstenergie des Elektrons', *Zeitschrift für Physik* **89**, 27–39.

(1939) 'On the self-energy of the electromagnetic field of the electron', *Physical Review* **56**, 72–85.

Weisskopf, V. F. (1983) 'Growing up with field theory: the development of quantum electrodynamics', in L. M. Brown and L. Hoddeson (eds.) *The Birth of Particle Physics* (Cambridge: Cambridge University Press), 56–81.

Weyl, H. (1928) *Gruppentheorie und Quantenmechanik* (Leipzig: Hirzel). English translation (1930) *The Theory of Groups and Quantum Mechanics* (New York: Dover).

(1929) 'Elektrons und Gravitation. I', *Zeitschrift für Physik* **56**, 330–52.
White, H. E. (1934) *Introduction to Atomic Spectra* (New York, London: McGraw-Hill Book Company).
Wigner, E. (1927) 'Einige Folgerungen aus der Schrödingerschen Theorie für die Termstrukturen' *Zeitschrift für Physik* **43**, 624–52.
Wilczek, F. (1990) *Fractional Statistics and Anyon Superconductivity* (Singapore: World Scientific).
Yang, C. N. and Mills, R. (1954) 'Conservation of isotopic spin and isotopic gauge invariance', *Physical Review* **96**, 191–5.
Zeeman, P. (1896) 'Over den invloed eener magnetisatie op den aard van het door een stof uitgezonden licht', *Verlag van de Gewone Vergaderingen der Wis-en Natur-kundige Afdeeling, Koninklijke Akademie van Wetenschappen te Amsterdam* **5**, 181–5, 242–8. English translation (1897) 'On the influence of magnetism on the nature of the light emitted by a substance', *Philosophical Magazine* **43**, 226–39.
Zweig, G. (1964a) 'An $SU_3$ model for strong interaction symmetry and its breaking', CERN preprint 8182/TH401 (17 January 1964).
  (1964b) 'An $SU_3$ model for strong interaction symmetry and its breaking: II', CERN preprint 8419/TH412 (21 February 1964).

# Index

Abelard, P., 92
adiabatic invariants, 41
Adler, S., 168, 169
Adler–Bell–Jackiw anomaly, 167–9, 174, 181
ADONE, 170, 171
Alembert, J. Le Rond d', 110
alkalis, 35, 43, 46, 74
  alkali doublets, 35, 43–4, 49, 53, 56, 59, 64, 79, 105, 106, 108, 130
alkaline earths, 35, 43, 73
analytic/synthetic distinction, 19
Anderson, C. D., 132
anomalous phenomena *see* spectroscopy
anticommutation relations, 112, 126, 127, 139–41, 155
anticommutator, 140
antiparticles, 131–3
antiquarks, 151
antirealism, 5, 186
antisymmetric function, 118, 119, 153, 163
antisymmetric state, 146, 155, 157
antitaxonomies, 89, 90, 92
anyons, 141
Appelquist, T., 171
Aristotle, theory of predicables, 91
asymptotic freedom, 164, 167, 169, 182
atomic core model, 35, 44–6, 50–2, 54, 55, 66, 70, 78, 79, 103, 106
  referred to as *Ersatzmodell*, 63
atomic core, 40, 45, 53, 56, 59, 65, 69, 71, 72, 80, 97, 101, 103
  angular momentum, 45, 50
  magnetic anomaly, 51, 58, 60, 64, 67, 102, 120, 130
atomic number, 39, 44, 68, 69, 158

*Aufbauprinzip see* Bohr, building-up principle
*Ausschließungsregel see* exclusion principle
axial-vector currents, 168

Back, E., 74
Bacry, H., 163
Balmer, J., 113, 151
Barnes, V. E., 151
baryons, 146, 151, 152, 164
  baryon spectra, 146, 153, 163, 172
  baryon number, 148, 149, 152
Beck, E., 51
Belinfante, F. J., 139
Bell, J. S., 168
$\beta$ decay, 133, 135, 146, *see also* Fermi
Bjorken, J. D., 165
Bjorken scaling, 166
Blackett, P. M. S., 133
Bloch, F., 134
Boethius, S., 91
Bohr, N., 12, 13, 20, 32, 35, 45, 56, 59, 62, 65, 67, 73, 75, 76, 80
  atomic theory, 36–7, 151
  Bohr–Sommerfeld quantum conditions, 37, 49, 51, 56, 59, 105, 106, 108
  building-up principle, 40–1, 46, 49, 51, 55, 59, 61, 62, 64, 71, 105, 106, 108
  building-up schema, 39–42, 71
  correspondence principle, 13, 40–1, 52, 53, 63, 73, 75, 80
  quantity $d$, 53–6, 97, 108
  non-mechanical constraint, 52–5, 71, 106, 114
Bohr–Coster theory, 70
Bohr–Kramers–Slater theory, 63–4

204

Bohr magneton, 38, 75, 128
Boltzmann, L. 41, 104
Born, M., 112, 113, 122, 126
Bose, S. N., 116
bosons, 140, 141, 155
   intermediate vector, 167, 168
Brahe, T., 20
Brookhaven National Laboratory, 151, 171
Buchdahl, G., 2, 23, 24, 26, 31, 142, 187

Cabibbo, N., 152, 172
Cabibbo angle, 152
Cao, T. Y., 187
Cartan, E., 16
Cassidy, D., 57
Cassirer, E., 2, 4, 7, 24, 28, 31, 33, 110, 142, 185–6, 187
   *Determinism and Indeterminism in Modern Physics*, 29–31
   *Substance and Function*, 28, 29
   *The Philosophy of the Enlightenment*, 29
   on scientific principles, 30
Catalan, M. A., 43
Cambridge Electron Accelerator, 170, 171
Cavendish Laboratory, 133
CDHS experiment, 166
centre-of-mass energy, 170, 171
Chadwick, J., 134
charm, 153, 163, 171, 172, 182
charmonium, 171, 182
Clausius, R. J. E., 41
Close, F. E., 164
colour, 145, 156
colour confinement, 153, 164
colour SU(3) symmetry, 156, 162, 163
coloured quark model, 165, *see also* quarks, coloured
commutation relations, 113, 120, 125, 126, 127, 155
commutator, 138, 140
conservation of charge, 148, 167
constitutive principles, 22, 23
   and Reichenbach, H., 2, 14–15
   constitutive a priori principles *see* Friedman, and relativized a priori principles
   *see also* regulative principles
Copernicus, N., 20
core magnetic anomaly *see* atomic core
Coster, D., 42
coupling

core–electron, 45, 53, 54, 60, 71, 106
*jj*-coupling, 66
   Russell–Saunders coupling, 66
   spin–orbit, 43–5, 57, 66, 107–9, 120, 121
coupling constant for QCD, 164, 166
Cowan, C. L., 134, 146
Curie, M., 11
current algebra, 152

Darwin, C. G., 120–2, 128, 130
Dauvillier, A., 44
de Broglie, L., 44, 113
deep inelastic lepton–hadron scattering, 165, 166, 182
de Groot, J. G. H., 166
Deilamian, K., 158
demonstrative induction, 103–7, 109, 173
Descartes, R., 29, 103
Dirac, P. A. M., 8, 33, 77, 113–18, 123–6, 128–32, 135, 136, 139
   equation for the electron, 32, 51, 77, 80, 112–14, 120, 128, 130, 131, 137, 143
   hole theory, 112, 114–15, 132–6, 138, 139, 140, 143
   'subtraction physics', 135–7, 139
Dirac spinor, 129
dispersion theory, 79
Dorling, J., 104
doublet riddle, 44, 46, 49, 51, 59, 62, 105, 108
Drake, G. W. F., 158
Duhem, P., 172, 173
Duhem–Quine thesis, 4, 34, 145, 147, 172–4
dynamic Kantianism
   and Friedman, M., 2, 7, 15, 18, 20–1, 24, 84
   and Kuhn *see* Kuhn, post-Darwinian Kantianism
   and Massimi, M., 2, 4, 7, 21, 24

Eco, U., 92
Ehrenfest, P., 41, 120
eightfold way, 145, 146, 150–1
Einstein, A., 16, 20, 30, 41, 116
electron
   charge-to-mass ratio, 38, 47
   orbital angular momentum, 38, 43, 45, 55, 57, 66, 70, 98, 108, 109, 130
   orbital magnetic moment, 38
   orbital magneto-mechanical ratio, 38, 51

electron (cont.)
  spin, 4, 8, 32, 35, 39, 55, 56, 65, 75, 77, 78, 80, 97, 98, 106–9, 112, 143, 151, 184
    two-valuedness of, 70
  spin angular momentum, 43, 45, 55, 66, 97, 108, 114, 120, 121, 130
  spin magnetic moment, 43, 51, 57, 58, 75–7, 107–9, 112, 114, 120, 121, 130
  spin magneto-mechanical ratio, 51
electron–positron annihilation, 148, 170, 171
electron–positron creation, 148
electroweak currents, 168
electroweak theory, 181
Elsasser, W., 133
Enlightenment, 110
Enz, C., 114
Eötvös, L. von, 15, 16, 20
epistemological holism *see* Quine
ETH, 138
Euclidean geometry, 15
exclusion principle, 1, 2, 4, 12, 13, 21, 24, 112, 145–7, 153–6, 162, 164, 177, 181, 182, 184, 187
  as antisymmetrization, 3, 8, 33, 116, 118, 141, 145, 185
  as phenomenological rule, 1–4, 7–8, 32, 33, 35, 42, 53, 65, 74, 75, 78–81, 103, 109, 110, 112, 113, 115–16, 120, 123, 141, 142, 143, 145, 184, 185
    *see also* Pauli
  as scientific principle, 1–4, 33, 72, 112, 114–15, 126, 127, 131, 138–4, 146
  empirical support for, 3, 4, 147, 181–2, 188
  experimental validation, 2, 3, 20, 145, 147, 157–62, 172, 174–80, 185
  nomological scope and strength, 3, 4, 7, 34, 114, 142, 145–6, 173, 174, 175, 183–5
  referred to as *Äquivalenzregel*, 119, 122
  referred to as *Ausschließungsregel*, 1, 73, 112, 114, 115
  referred to as Pauli's *Verbot*, 1, 33, 73, 118, 126, 142, 146
  regulative function of 2–4, 7, 31–2, 185

Fermi, E., 8, 33, 114, 116, 118, 119, 134, 149, 171
  theory of $\beta$ decay, 134, 143, 168
Fermilab, 160
fermions, 8, 126, 140, 141, 153, 155, 157

Feynman, R., 152, 165
Fierz, M., 138–9, 140
fine structure constant, 68, 135
Fitzgerald, G., 12
Forman, P., 44, 45, 49
Fowler, A., 72
French, S., 5, 154, 155, 173, 178, 187
Frenkel, J., 76
Fresnel, A., 10, 12
Friedman, M., 2, 14–19, 22, 32, 79, 142
  *see also* dynamic Kantianism
  and relativized a priori principles, 7, 15–20, 24, 79, 84, 142
  on Kant, 22–3, 24
Fritzsch, H., 153
Frobenius, G. F., 123

Galilei, G., 20
Galison, P., 174
$\gamma$-ray scattering, 148
gauge invariance, 137, 167–8
gauge theory, 167
Gell-Mann, M., 145, 146, 149–53, 163, 164, 169
generalized rays, 155, 177, 178
Gentile, G., 154
Gillaspy, J. D., 158
Glashow, S. L., 167
Glashow–Iliopoulos–Maiani theory, 171, 181
gluons, 163–6
Goldhaber, M., 157, 159, 160, 171
Gordon, W., 129, 130
Goudsmit, S., 32, 75, 76, 80, 99, 113, 114, 119, 128
Green, H. S., 155, 156, 169
Greenberg, O. W., 147, 154–60, 162, 169, 177
Gross, D., 164
group theory, 115, 122–3, 148
Guyer, P., 26–7

ter Haar, D., 154
Hacking, I., 87, 89–93, 166
  and Kuhn's untranslatability thesis, 88–9
hadronic weak currents, 152
hadrons, 1, 146, 148, 151, 152, 163, 164, 165
Han, M. Y., 163, 169
Harper, W., 104
Heilbron, J. L., 44, 72, 74

Heisenberg, W., 1, 32, 35, 45, 54, 55, 61, 62, 65, 67, 73, 75, 76, 79, 98, 112, 113, 118–21, 123, 133, 135–7, 148
 first core model, 55–9, 64, 66
 new quantum principle, 62–4, 67, 99, 101
 second core model, 62–5, 106
 sharing principle, 55, 56, 59, 97, 106, 108, 114
helium atom, 54, 79, 118, 119, 123, 158–9
Helmoltz, H., 41
Hertz, G., 41
Hesse, M., 87
Hevesy, G., 42
Higgs, P. W., 167
Hilbert, D., 122
Hilborn, R. C., 146
Hofmann, W., 174
holes, 131–2, 134, 137
Husserl, E., 186
hydrogen atom, 113
hyperons, 146–8, 149, 152

incommensurability, 4, 17, 20, 33, 78, 80–2, 94, 97, 99, 102, 143
 as untranslatability between lexicons, 4, 78, 82–90, 94, 97, 103
  Kuhn's argument for untranslatability, 4, 86–90, 92–4
indistinguishability postulate, 154, 155, 176
indistinguishable particles, 5, 8
Institute for Advanced Study, 140
interaction
 electromagnetic, 152, 167
 electroweak, 167
 strong, 148, 167
 weak, 148, 152, 167
isospin, 148–9, 150

Jackiw, R., 168
Jammer, M., 41
jets, 170, 174
Jordan, P., 112–14, 120, 121, 126–8
Joyce, J., 151
J/psi particle, 171, 182

Kamefuchi, S., 155
Kant, I., 2, 7, 14, 15, 22–4, 29, 31, 95–6, 187
 and the a priori, 7, 14
 and the conception of lawlikeness, 22–4

 and the logical principle of continuity, 25, 95
 and the logical principle of homogeneity, 25, 95
 and the logical principle of specification, 25, 95
 *Critique of Pure Reason*, 21–3, 25
 *Critique of the Power of Judgment*, 22, 25, 26, 95
 *Metaphysical Foundations of Natural Science*, 22
 *Opus postumum*, 22, 24
 *Prolegomena*, 22
kaons, 147–8, 151, 153
Kayser, H., 52
Kepler, J., 20, 22
Kitcher, P., 27, 87
Klein, O., 129, 130
Klein–Gordon equation, 129, 130, 134, 136–8
Klein–Nishina formula, 130
Klein paradox, 130
Kragh, H., 42
Kramers, H. A., 79
Kripke, S., 89
Kronig, R., 74–5
K-shell, 69
 X-rays, 157, 159, 160
Kuhn, T. S., 4, 13, 20, 78–86, 89–90, 92–4, 96, 97, 99, 102, 131, 132
 and no-overlap principle, 82–3, 86, 87
 and nomic generalizations, 83, 103, 110, 143
 and normic generalizations, 83
 problem of the new world, 87–8, 93, 94, 96, 97
 post-Darwinian Kantianism, 84, 93–5, 97, 98
 *see also* incommensurability, lexical taxonomies
 *The Structure of Scientific Revolutions*, 81
 *The Essential Tension*, 81

Ladyman, J., 187
Lakatos, I., 12–13
Landé, A., 7, 32, 35, 44–6, 48–57, 61, 66–70, 74, 79, 98, 99, 115, 120
 Landé g factors, 49–50, 57, 60, 63, 64, 66, 75, 98, 105, 107, 109, 114, 123, 130, 151
Landé–Heisenberg branching rule, 61–4, 70, 71, 98–101, 106, 108, 114

Larmor's theorem, 38, 59, 67, 80, 101, 105–9
  and failure, 51, 58, 65, 67–71, 120
  Larmor frequency, 38, 47, 49, 57, 67, 68
Laudan, L., 173, 180
lead, spectrum of, 74
Lederman, L. M., 146
Leibniz, G., 5, 29
leptons, 1, 165
LeRoy, E., 10
Lévy, M., 152
lexical taxonomies, 85–102
Lie algebra, 150
Lipton, P., 84
logical positivism, 13–14
Lorentz, H. A., 12
Lorentz invariance, 123, 130, 137
Lorentz normal triplet, 48
Lorentz unit of the Zeeman effect, 48

Macrae, K. I., 156, 169
Majorana, E., 139
Massimi, M., 5, 105, 141, 166, 173
matrix mechanics, 79, 112, 113, 116, 120, 122
Maxwell, J. C., 10, 15
  Maxwell's equation, 132, 135
Maxwell–Boltzmann statistics, 116, 118, 176
Mehra, J., 113
Mendeleev, D., 1, 39, 46, 72, 158
mesons, 146, 151, 152, 164
Messiah, A. M. L., 147, 154–6, 169, 177
metaphysical realism, 186, 187
Michelson–Morley experiment, 12, 15–16
microcausality, 138–41
Mill, J. S., 88, 90
Mills, R., 167
Minkowski, H., 15
Mohapatra, R. N., 157–60
multiplets, 148, 150
muons, 115, 153, 165

Nambu, Y., 163, 169
Ne'eman, Y., 149–151
negative energy sea, 131–2, 134
Nelson, C. A., 156
neon, spectrum of, 61
neutral pion, 168–9
  decay, 168, 174, 181
neutrino, 134, 143, 146, 165
neutron, 115, 133, 134, 146, 148, 149, 152, 165

Newton, I., 20, 29, 109, 110
  law of gravitation, 10, 14, 22, 23, 84, 103
  method of deduction from phenomena, 4, 22, 29, 33, 103–4, 105, 110
  Newtonian mechanics, 14, 15, 22, 23, 83, 84
  principles, 14, 19, 22, 23, 28
  second law, 83–4
  *Principia*, 29, 103, 104
  *Optics*, 29, 104
  *System of the World*, 104
Nickles, T., 102
Noether, E., 122
nominalism, 93
Norton, J., 104, 105

objectivity, 186, 187
Occhialini, G. P. S., 133
octet of spin-1/2 baryons, 149, 152
Ohnuki, Y., 155
Okun, L. B., 159
$\Omega^-$ particle, 145, 146, 148, 150–3, 162, 172
$\Omega^-$ decuplet, 149, 151–3
operator
  colour, 163
  creation and annihilation, 125
  occupation number, 125, 126
  permutation, 155
Oppenheimer, J. R., 132, 136
optical doublets *see* alkali doublets

parafield theory, 155, 156
paraparticles, 144, 155
parastatistics, 2, 3, 8, 145, 147, 154–7, 169, 172–6, 185
  gauging, 156
  para-Bose, 34, 156, 176, 177
  para-Fermi, 34, 156, 176, 177
parity, 149
  violation, 168
parons, 34, 157–62, 174, 178, 188
parton model, 165–7, 169, 182
partons, 165, 182
Paschen–Back effect, 35, 52, 57, 58, 65, 71, 72, 99, 105–7, 109, 130
Paschos, E. A., 165
Pauli, W., 1, 7, 8, 32, 35, 40, 42, 45, 59, 60, 65–7, 70, 74–7, 80, 98–100, 107, 109, 112–5, 119–22, 128, 130, 133–40, 143, 146, 154

and the relativistic correction in the Zeeman effect, 67–70, 99, 101
and the exclusion rule, 72–3
and the electron's *Zweideutigkeit*, 32, 35, 60–2, 70–5, 77, 78, 80–1, 99, 101–10, 112, 114, 130
and *zweideutig* quantum numbers, 63, 64, 98, 99, 100, 110
Pauli spin matrices, 112, 121–3, 129, 141, 142
Pauli–Weisskopf 'anti-Dirac paper', 115, 133, 135–8
Peierls, R. E., 133
periodic table, 1, 39, 40, 42, 46, 71–3, 109
permanence of the $g$-sum rule, 66–7, 72, 73, 99, 107, 110, 114
permutation invariance, 3, 34, 176–9
perturbation theory, 116, 164
Petruccioli, S., 40, 64
photons, 167
Pickering, A., 151, 173
Planck, M., 20, 37
Plesset, M. S., 136
Poincaré, H., 9–11
  and conventionalism, 7, 10–11, 14
  and structural realism, 9–10
Politzer, H. D., 164, 171
Popper, K., 11–13
Porphyry, 91, 92
  taxonomic tree, 91–2
positron, 112, 115, 132–4, 137, 143, 146, 187
postulate of positive energy, 139, 140
Preston, T., 48
prospective intelligibility of scientific revolutions, 82, 102
  *see also* revolutionary transition around 1924
protons, 115, 131–3, 146, 148, 149, 151, 152, 165
Putnam, H., 89, 186–7

$q$-numbers, 124, 125
quanta, 125
quantum chromodynamics, 2, 3, 8, 34, 115, 145, 147, 156, 164, 166, 167, 169, 172–4, 181, 185
quantum electrodynamics, 115, 124, 152, 164, 167
quantum field theory, 115, 125, 126, 134, 141, 142, 155, 167
quantum mechanics, 2

non-relativistic, 115, 119, 142
relativistic, 115, 128, 143
quantum number, 8
  azimuthal quantum number, 8, 37, 39, 43, 45, 46, 56, 65, 70
  core quantum number, 45, 46, 50
  inner quantum number, 44–6, 49, 54, 72
  magnetic quantum number, 8, 38, 45, 49, 54, 57, 65, 66, 70, 72
  principal quantum number, 8, 37, 39, 43, 70, 109
  radial quantum number, 37
quantum statistics, 115–18
  Bose–Einstein statistics, 34, 118, 124–7, 133, 137–41, 154, 155, 162, 176, 177
  Fermi–Dirac statistics, 3, 8, 21, 33, 112, 119, 126, 127, 133, 139, 141, 142, 153–5, 162, 177, 184
quantum theory, 80
  old quantum theory, 1–4, 32, 35, 38, 51, 78–80, 97, 102, 103, 106, 107, 109, 113, 114, 130, 143, 184
  and theoretical assumptions, 3, 4, 33, 80, 105, 106, 109, 143, 184
  new quantum theory, 2–4, 35, 51, 79, 80, 97, 106, 109, 113, 124, 142, 143, 146
quarks, 1, 8, 146, 151–4
  as parafermions, 154, 156, 162
  coloured quarks, 3, 144, 145, 147, 162–72, 173, 182, 185, 187
  flavours, 151, 153, 162
  quark theory, 4, 34, 145–7, 153, 154, 156, 162, 172
Quine, W. V. O., 13, 19, 85, 172, 173, 181
  epistemological holism, 4, 18, 34, 147, 172–4, 179–81
  *see also* underdetermination
quons, 162, 178

Ramberg, E., 159–61
ratio of hadron-to-muon production $R$, 170–2, 174
rationality
  prospective, 16, 17, 32
  retrospective, 16
Rechenberg, H., 113
Redhead, M., 5, 141, 155, 178
regulative function, 3, 5
  and Cassirer, 28
  as opposed to constitutive function, 2, 4, 142
  *see also* exclusion principle

regulative principles, 22, 23, 95
Reichenbach, H., 2, 14–15, 142
   and axioms of coordination, 7, 14, 79
   *The Theory of Relativity and A Priori Knowledge*, 14
Reines, F., 134, 146
relativistic doublets *see* X-ray doublets
relativity theory
   and the equivalence principle, 3, 15–17, 30, 79
   and the light principle, 3, 15, 16, 79
   general, 3, 14–16, 20
   special, 3, 15, 16, 76, 77, 109, 140
renormalizability, 167, 168, 181
resonances, 149–50, 153, 171
revolutionary transition around 1924, 4, 35, 78–80, 98, 100, 184
   prospective intelligibility of, 4, 103, 106, 109, 143, 184
   retrospective intelligibility of, 143
Richter, B., 171
Rickles, D., 178
Rubinowicz, A., 45, 58–9
Runge, C. D., 48
Runge fractions, 48
Rutherford, E., 71
Rydberg, J., 36, 37, 72, 107
   constant, 109
   rule for closure of electronic groups, 110

Salam, A., 167
scaling variable, 166
scaling violations, 164–6, 169, 174, 182
Scharff-Goldhaber, G., 157, 159
Schrödinger, E., 79, 113
Schrödinger equation, 124
Schur, I., 123
Schwartz, M., 146
scientific lexicon, 82–7, 102, 111
scientific realism, 5, 184, 186
second quantization, 125, 127, 138, 155
self-energy of the electron, 135, 136, 139
Serwer, D., 41, 46, 62
Silvestrini, V., 170
SLAC, 166
Slater, J. C., 123
Smith, G. E., 104
Snow, G., 159–61
SO(3), 156
sodium principal doublet, 48, 49

Sommerfeld, A., 13, 20, 37–9, 44, 45, 56, 57, 72, 130
   and space quantization, 39, 45, 59, 105–7, 109
Sommerfeld–Voigt formula, 57
SPEAR, 171
spectroscopy, 3
   anomalous phenomena, 3, 4, 32, 35, 44, 70, 80, 99, 103, 105, 106, 109, 112, 130, 184
spin *see* electron
spin–relativity doublets, 108, 109
spin–statistics theorem, 2, 3, 8, 21, 33, 112, 115, 134, 137–42, 144, 146, 147, 153, 154, 156, 172, 184
spinning electron model, 32, 57, 75–7, 79, 80, 108, 113, 114, 119, 120, 128, 130, 142, 143
Spinoza, B., 29
statistical weights, 41, 46
   invariance and permanence of, 41
   riddle of, 46, 49, 51, 59, 60–2, 64, 71, 105, 108
Steinberger, J., 146
Stern–Gerlach experiment, 39, 107
Stolt, R. H., 155
Stoner, E., 42, 71–2, 73
   Stoner's rule, 107, 110, 114
strangeness, 148–50, 152, 153
   conservation, 152
strangeness-changing weak neutral currents, 153, 172, 182
structural realism, 187
SU(2), 148
SU(3), 150, 152
supermultiplets, 149–51
surplus structure, 178–9
symmetric function, 118, 119, 125, 153
symmetric state, 146, 155, 157
symmetrization postulate, 141, 154, 155, 176
symmetry breaking
   anomalous or quantum mechanical, 168
   spontaneous, 167
systematic unity, 3, 25–7, 30, 95, 96, 185, 186
systematization, 142

Taylor, J. R., 155
Teller, P., 5, 155
Thomas, L. H., 76

Thomas's relativistic correction factor, 113, 119, 122, 128, 130, 143
't Hooft, G., 167
Thomson, J. J., 38
Ting, S., 171
Tino, G. M., 162
Tomonaga, S., 125
total angular momentum, 43, 45, 50, 54, 55, 57, 59, 61, 66, 98, 100, 108, 109
transformation theory, 128
truth, 186–7

Uhlenbeck, G. E., 32, 75, 76, 80, 99, 113, 114, 119, 128
underdetermination, 4, 145, 147, 173, 180
   Quinean, 147, 148
unitary spin, 150

vacuum polarization, 135, 136
Van der Waerden, B. L., 76, 122
van Fraassen, B., 186
   and objectifying inquiry, 186
van Vleck, J. H., 54
vector currents, 168
vector model, 43
Voigt, W., 57
Von Neumann, J., 112, 123

Ward, J. C., 167
wave mechanics, 79, 113, 116, 120
Weinberg, S., 141, 167
Weisskopf, V., 133, 134, 136, 139
Wentzel, G., 119
Weyl, H., 122, 131, 132
Wigner, E., 112, 114, 122, 123, 126, 127
Wilczek, F., 141, 164
Wittgenstein, L., 183

X-ray doublets, 44, 49, 59, 60, 64, 79, 105, 106, 108, 131

Yang, C. G., 167
Young pattern, 157–8

Zeeman, P., 47, 48
Zeeman effect
   anomalous, 35, 48, 52–5, 57, 58, 60, 64, 65, 70, 74, 79, 99, 105–8, 120, 130, 143
   *mg* splitting factors, 49–50
   normal, 48
   Zeeman Hamiltonian, 49
*Zwang see* Bohr, non-mechanical constraint
*Zweideutigkeit see* Pauli
Zweig, G., 151, 153